Serious Managers Guide to AI Governance

Navigating the Future of AI-based Intelligent Oversight

Claude Louis-Charles, Phd

1

Cybersoft Publishing LLC

Fort Washington, MD 20744

First Edition April 2026

Table of Contents

Introduction

Most organizations didn't decide to become "AI organizations." It crept up on them. A pilot chatbot here, an analytics model there, a vendor tool with a recommendation engine under the hood. Then one day a senior leader asks a simple question you're expected to answer: "Are we sure this thing is safe?" That question is why this book exists.

1 The New Era of AI Governance

1.1 Opening vignette: When "Innovation First" Backfires

On Monday morning, the CIO of a large public sector agency walked into an executive briefing expecting to celebrate a successful AI pilot. The team had deployed a new machine learning model to prioritize citizen service requests, promising faster response times and better use of limited resources. The dashboard showed impressive numbers: reduced backlog, shorter response times, and lower call volumes to the contact center. On the surface, the AI initiative looked like a textbook modernization win.

By Wednesday, the story had changed. Frontline staff began reporting that certain neighborhoods were no longer seeing their cases prioritized, even though their requests were urgent. A community organization flagged that the system seemed to be deprioritizing requests from older citizens and non-native English speakers. Social media picked up the complaints, local news followed, and the agency's leadership suddenly faced questions not just about performance, but fairness, transparency, and accountability. No one could clearly explain how the model made its decisions, why it behaved this way, or who had approved it to go live with real citizens.

By Friday, the AI system was abruptly switched off. The organization reverted to manual triage, losing many of the efficiency gains it had briefly enjoyed. Executives ordered an internal review, regulators started asking questions, and the CIO's team found themselves in a familiar but painful position: the technology had worked as designed, but the governance around it had not. There had been no clear definition of acceptable risk, no agreed thresholds for fairness or explainability, and no structured oversight to evaluate whether the model's behavior aligned with the organization's mission and values.

This kind of scenario is no longer hypothetical. It is becoming common across sectors: healthcare providers whose diagnostic tools perform differently across demographic groups, banks whose automated decisions raise fairness concerns, and government agencies whose AI systems amplify existing inequities. The technology is powerful, but without a disciplined approach to AI governance, the risk of misalignment, reputational damage, and regulatory exposure grows with every new deployment. That is the world this chapter prepares you to navigate.

1.2 Why This Chapter Matters for Managers

If you are an IT manager, AI is no longer "someone else's problem." Even if you are not directly responsible for data science or machine learning, AI is showing up in the systems you buy, the platforms you integrate, and the services you support. Vendors are embedding models in their products. Business units are experimenting with generative AI tools. Shadow IT is creating its own automations. In this environment, governance is not a luxury; it is how you keep control of your environment while enabling innovation.

AI governance is fundamentally about decision-making under uncertainty. It asks who can use AI, for what purposes, with what safeguards, and under whose oversight. It forces clarity about acceptable risk, lines of accountability, and mechanisms for monitoring and improvement over time. Without that clarity, you get fragmented experiments, inconsistent controls, and surprise consequences. With it, you can say "yes" to AI more often, because you have a way to ensure it stays aligned with your mission and obligations.

This chapter sets the foundation for the entire book. It gives you language to explain AI governance to senior leaders and peers, defines the problem space in manager-ready terms, and introduces the core dimensions you will use throughout the rest of the chapters. By the end of this chapter, you should be able to articulate why AI governance is different from traditional IT governance, what is at stake if you get it wrong, and how to frame AI governance as an enabler of modernization—not a brake on progress.

1.3 From "Systems of Record" to "Systems That Decide"

For most of the last few decades, IT governance has focused on systems of record. These systems stored information, processed transactions, and provided reports. The central governance questions were about reliability, security, availability, and compliance. Did the system stay online? Was the data accurate? Were access controls appropriate? Could you pass an audit? The decision-making authority remained largely with humans; the systems supported their work but did not make consequential choices on their own.

AI changes that equation in a fundamental way. Modern AI systems are not just systems of record; they are systems that predict, recommend, and increasingly decide. Instead of merely storing data, they infer patterns, classify people or events, and suggest or automate actions. This shift from "systems of record" to "systems that decide" creates a new category of governance questions: when is it acceptable for a system to make a decision, under what conditions, with what transparency, and how do you ensure that those decisions remain aligned with policy, law, and organizational values over time?

As AI becomes embedded across workflows—triaging tickets, suggesting diagnoses, scoring risk, or drafting communications—the distinction between "supporting" and "deciding" blurs. In many organizations, human oversight becomes nominal rather than substantive, especially under pressure to move faster and handle more volume.

Governance must therefore adapt. It must recognize that decisions can now be distributed between humans and algorithms, and it must regulate not only the system's technical behavior, but the human-system interaction as a whole.

1.4 The Convergence That Makes Governance Harder—and More Necessary

This new era is defined by a convergence of three forces: data at scale, automation at speed, and cloud everywhere. Each of these forces on its own is manageable; together, they create a landscape where small governance gaps can quickly lead to large consequences.

First, data at scale means that AI systems are trained and operate on vast volumes of information collected from many sources. Historical records, behavioral data, operational logs, and external datasets all feed into models. The more data is used, the harder it becomes to trace precisely which data influenced which model behavior and when. This makes questions about data quality, bias, and consent more complex than in traditional IT environments.

Second, automation at speed accelerates the impact of every decision. In a manual process, a flawed decision might affect dozens of cases before someone notices a pattern. In an AI-powered process, the same flaw could affect thousands of individuals in hours. Automation amplifies both benefits and harms. Governance must therefore aim not only to prevent errors but to contain and detect them quickly when they inevitably occur.

Third, cloud everywhere relocates where and how systems are run. Models may be trained in one environment, deployed in another, and consumed via APIs you do not fully control. Vendors may update models or infrastructure without your direct involvement. This increases the importance of understanding shared responsibility; certain controls remain with you, while others shift to providers. Governance must provide clarity on who is accountable for which aspects of an AI system's behavior in a multi-party ecosystem.

When these forces converge, traditional incremental governance approaches break down. Change review boards and one-time approvals cannot keep up with dynamic models that are retrained frequently, or systems that adapt to new data. You need governance that is continuous, integrated into the lifecycle of AI systems, and tightly linked to the organization's strategy and risk appetite. That is the core message of the new era of AI governance.

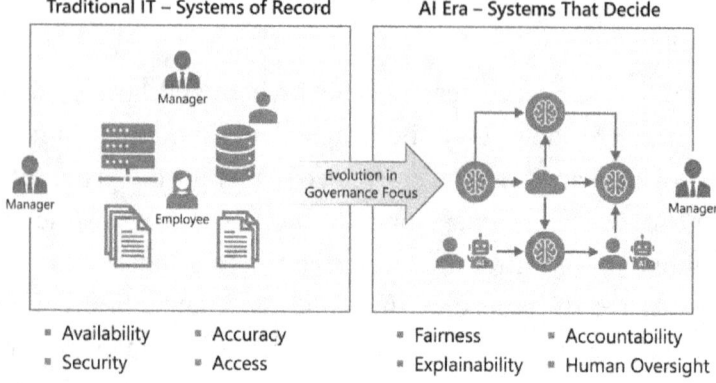

Traditional IT – Systems of Record		AI Era – Systems That Decide	
▪ Availability	▪ Accuracy	▪ Fairness	▪ Accountability
▪ Security	▪ Access	▪ Explainability	▪ Human Oversight

1.5 Four Defining Dimensions of Modern AI Governance

To make AI governance manageable, you need a simple structure—a way to categorize and discuss the challenges without drowning in detail. Throughout this book, we will use four defining dimensions of modern AI governance: value, risk, control, and trust. These dimensions give you a consistent lens to evaluate AI initiatives and guide your decision-making across different domains, vendors, and use cases.

Value is about why the AI exists at all. A governed AI system must have a clear mission-aligned purpose. It should support an outcome that matters: better citizen services, improved healthcare quality, more efficient operations, reduced fraud, or more accurate decisions. When the value is vague or undefined, governance drifts into bureaucracy. When value is explicit, governance can ensure that controls are proportionate to the importance of the outcome and the stakeholders involved.

Risk focuses on what could go wrong and for whom. AI introduces new kinds of risk—unintended bias, privacy violations, opaque decisions, security vulnerabilities related to data or models, and regulatory non-compliance. A practical governance approach does not try to eliminate risk; instead, it helps you understand which risks are acceptable, which require mitigation, and which must be avoided. It recognizes that risk tolerance may differ across use cases and stakeholder groups, and that high-risk use cases demand heavier oversight.

Control is about the mechanisms you establish to guide and constrain AI behavior. Controls include policies, procedures, technical safeguards, monitoring, and human oversight. Good governance does not rely on a single control; it layers them so that if one fails, others can detect and correct deviations. For AI, control mechanisms might include approval processes for new models, defined human-in-the-loop roles, constraints on model inputs and outputs, and thresholds that trigger model review or rollback.

Trust is the outcome you are ultimately trying to maintain— trust from leadership, staff, customers, regulators, and the public. Trust is not created by marketing or messaging; it is earned by consistent, transparent governance. When people believe that your AI systems are well-managed, fair, and accountable, they are more likely to adopt them, support their use, and grant you the benefit of the doubt when something goes wrong. In this sense, trust becomes a strategic asset, built slowly through good governance and lost quickly when governance fails.

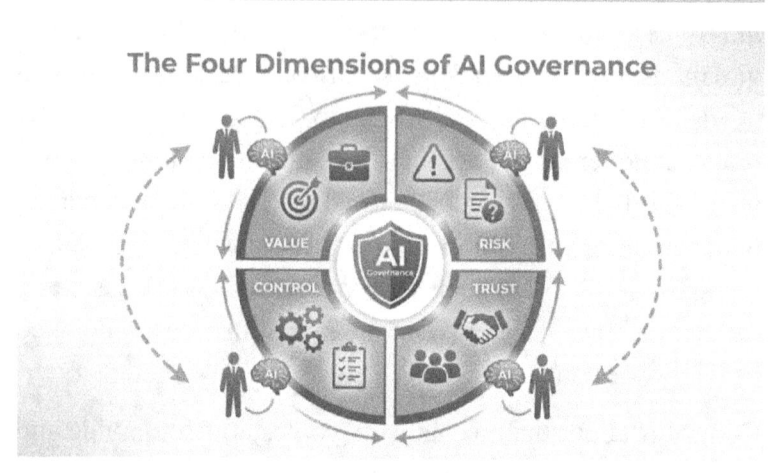

1.6 New Risks: Algorithmic Bias, Model Drift, and Opacity

Three concepts appear repeatedly in AI conversations and are central to why governance is so critical in this era: algorithmic bias, model drift, and opacity. As a manager, you do not need to become a data scientist. Still, you do need to understand these concepts at a practical level—enough to ask good questions, evaluate trade-offs, and design oversight that actually works in your organization.

Algorithmic bias refers to systematic and unfair differences in how AI systems treat individuals or groups. It can arise from biased historical data, incomplete datasets, flawed label definitions, or the way outcomes are optimized. For example, a model trained on historical hiring decisions may learn to replicate past discrimination, even if no explicit protected attributes are used. From a governance perspective, algorithmic bias is not only a technical issue; it is also a policy and ethical question that requires clarity on fairness goals, acceptable trade-offs, and methods for measurement and mitigation.

Model drift describes how an AI system's performance changes over time as the world around it evolves. A model built on last year's data may gradually become less accurate or less fair as population behaviors, economic conditions, or operational patterns shift. In some cases, drift is subtle and slow; in others, it is rapid, especially when external shocks occur. Without governance mechanisms for monitoring, testing, and recalibration, drift can erode the reliability and fairness of AI systems, often going unnoticed until damage is done.

Opacity is the challenge of understanding how an AI system arrives at its outputs. Some models are intrinsically complex, making it difficult to trace the path from input to decision. Even when explanations exist, they may not be comprehensible to non-technical stakeholders. Governance must therefore address not only whether a model can be explained technically, but whether the explanation is meaningful and appropriate for the audiences who depend on the system—executives, regulators, impacted individuals, and front-line staff.

1.7 The Manager's Reality: Pressure, Ambiguity, and Accountability

In practice, you operate in an environment filled with competing pressures. Business or mission leaders want to move fast and capture value. Vendors promise ready-to-use AI capabilities that sound compelling. Your teams are stretched across multiple priorities, from cybersecurity to cloud migration to legacy system maintenance. At the same time, regulators, auditors, and oversight bodies are becoming more aware of AI risks and more willing to scrutinize systems that affect citizens, customers, or employees.

This creates tension. On one side is the narrative of innovation—be bold, experiment, transform. On the other hand, there is the reality of accountability—you are responsible for the systems you introduce or approve, even if others are building them. AI governance is about navigating that tension. It gives you a structured way to say "yes, and" rather than "no" or "go ahead and hope for the best." It allows you to frame requests and proposals in terms of value, risk, and required controls instead of vague enthusiasm for AI.

Ambiguity is another part of your reality. Regulations are evolving, internal policies may not yet explicitly address AI, and best practices are still being solidified. You may have data scientists or rely entirely on vendors. You may have central governance bodies or highly decentralized decision-making. In all these cases, your role as an IT manager is to bring structure where there is none, to ask consistent questions, and to push toward governance that is responsible by design rather than reactive and ad hoc.

1.8 AI Reality Check: Governance Is Not a Blocker

A common misconception is that governance slows down AI adoption or stifles innovation. In many organizations, governance is associated with checklists, delays, and bureaucratic overhead. If that is the culture, it is understandable that teams look for ways around governance when they want to innovate quickly. However, in the new era of AI, treating governance as a brake is both inaccurate and dangerous.

Properly designed AI governance accelerates responsible innovation. It creates reusable patterns, pre-agreed risk thresholds, and standardized review processes. Instead of debating from scratch whether each new AI idea is allowed, you have a framework that defines categories of use, associated controls, and clear escalation paths. Teams know what is required for low-risk experiments versus high-impact deployments. This predictability reduces friction and shortens the time from idea to safe implementation.

The reality is that ungoverned AI eventually slows you down more than any governance process ever could, because you end up spending time on remediation, crisis management, and post-hoc explanations. A mature manager-ready governance approach reduces these downstream costs by building quality and responsibility into the AI lifecycle from the start. In that sense, governance is not a blocker; it is the infrastructure that enables scalable, sustainable AI modernization.

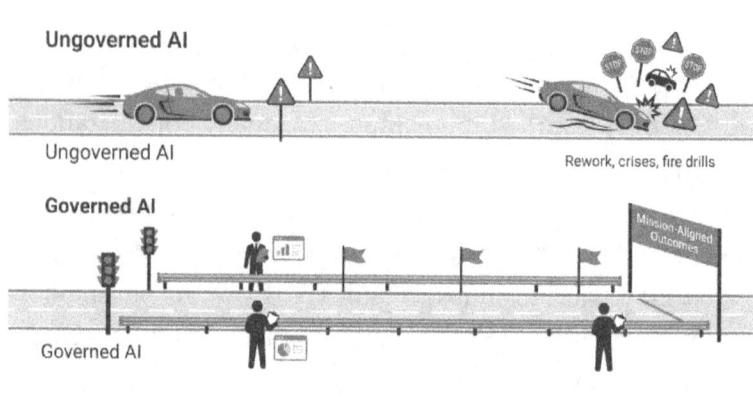

1.9 The Manager's Playbook: Questions to Ask on Monday Morning

To make this chapter immediately useful, here is a manager's playbook—questions you can start using on Monday morning when AI enters the conversation. These questions do not require you to be a technical expert; they require you to be a disciplined leader who insists on clarity.

- What mission or business outcome are we trying to improve with this AI?
  This clarifies the value and keeps the initiative anchored in measurable, mission-aligned goals rather than abstract enthusiasm for technology.

- Who will be affected by the AI system's decisions or predictions?
  This grounds the conversation in real stakeholders and helps surface potential fairness, equity, and reputational concerns early.

- What decisions will the AI system make or influence, and what remains in human control?
  This reveals the degree of autonomy, highlights where human oversight is needed, and tests whether that oversight is substantive or merely symbolic.

- What data will this AI use, and where does that data come from?
  This prompts attention to data quality, relevance, provenance, and any consent or privacy implications that might be overlooked.

- How will we know if the AI system is performing as intended over time?
  This forces the team to define metrics, monitoring plans, and triggers for review or rollback, guarding against silent model drift and degradation.

- Who is accountable if the AI system behaves in a way that conflicts with policy or harms stakeholders?
  This clarifies ownership, makes responsibility explicit, and reduces the risk of "the system" becoming a convenient but unacceptable scapegoat.

These questions are not exhaustive, but they create a baseline of governance thinking. They shift conversations from "Can we build it?" to "Should we build it, and under what conditions?" As you progress through this book, you will expand and refine this playbook with more detailed frameworks and templates.

1.10 Responsible AI Lens: Governance as Safety Net and Compass

Responsible AI is often described in terms of high-level principles, such as fairness, transparency, privacy, accountability, and human-centric design. These principles are important, but they can feel abstract and distant from daily work. AI governance is the process of translating those principles into concrete decisions about systems, processes, and behaviors. It serves as both a safety net and a compass for your AI efforts.

As a safety net, governance catches failures before they cause systemic harm. It ensures that models are tested for bias, that sensitive data is protected, that fallback options are in place when systems fail, and that incident response plans are in place. It encourages continuous monitoring rather than one-time approval. As a compass, governance helps you steer AI initiatives toward outcomes that align with your organization's mission and values. It guides trade-offs—for example, between efficiency and fairness—based on shared principles rather than ad hoc judgments.

By viewing every AI initiative through a Responsible AI lens, you make it clear that success is not only about performance metrics, but also about how those metrics are achieved and who is impacted. That perspective becomes integral to decision-making, procurement, design, and operations. Over time, it shapes culture, expectations, and trust, making responsible behavior the default rather than the exception.

1.11 The Emerging Regulatory and Stakeholder Landscape

In this new era, external expectations are changing rapidly. Regulators are increasingly attentive to how organizations use AI, particularly in high-stakes domains such as finance, healthcare, employment, and public services. New and evolving laws and guidelines focus on transparency, accountability, and risk management. Even where regulations are not yet specific or fully enforced, there is a clear direction of travel: organizations will be expected to demonstrate control over their AI systems and to explain their decisions.

Beyond formal regulation, stakeholders such as customers, citizens, employees, and advocacy groups are demanding more visibility into how AI affects them. Questions like "Why was I denied this service?" or "Why did this system flag me as high-risk?" are becoming more common. If your organization cannot answer these questions coherently, trust erodes. Media coverage of AI incidents can amplify even small missteps, turning technical failures into reputational crises that reach senior leadership quickly.

For managers, this means that AI governance cannot be treated as a narrow compliance exercise. It must be designed with a broad range of stakeholders in mind. It must anticipate questions from oversight bodies and the public. It must prepare you to provide evidence of how your AI systems are designed, tested, deployed, and monitored. This book will not give you legal advice, but it will help you build governance capabilities that are robust, auditable, and adaptable as the regulatory landscape evolves.

1.12 How This Chapter Connects to the Rest of the Book

This chapter has introduced the core idea that we are entering a new era of AI governance, shaped by systems that decide, the convergence of data, automation, and cloud, and the growing importance of risk, control, and trust. It has given you a manager-ready vocabulary for discussing AI governance and a starting playbook of questions to use immediately. The remaining chapters build on this foundation and turn these ideas into concrete structures and practices that you can put to work in your organization.

In Chapter 2, we will draw a clear line from traditional IT governance to AI governance, showing what stays the same, what changes, and how to integrate AI considerations into existing governance bodies and processes. You will see how familiar concepts such as change management, risk registers, and architecture reviews can be extended to address AI-specific challenges without reinventing your entire governance structure. You will also learn how to avoid creating a separate, siloed AI governance regime that conflicts with or duplicates your current practices.

As you move through the book, you will learn to design an AI operating model, define roles and responsibilities, operationalize ethics, govern data for AI, manage risk and compliance, and build a governance-aware culture. Each chapter will add tools to your manager's playbook, from checklists and templates to models and metrics. By the end, you will have a practical, mission-aligned approach to AI governance that you can adapt to your context and scale over time.

1.13 Key Takeaways from Chapter 1

- AI governance is about managing systems that decide, not just systems that store and process data. This shift introduces new questions about autonomy, fairness, and accountability that traditional IT governance alone cannot address.

- The convergence of data at scale, automation at speed, and cloud everywhere makes governance more complex and more critical. Small design decisions can now have large, rapid, and far-reaching consequences across stakeholders and services.

- Four dimensions—value, risk, control, and trust—provide a simple, repeatable lens for evaluating AI initiatives.
  They help you balance innovation with responsibility and align AI projects with mission and regulatory expectations.

- Algorithmic bias, model drift, and opacity are central governance challenges, not niche technical details. Managers must understand them well enough to ask informed questions, demand evidence, and design oversight.

- Governance, when designed well, accelerates responsible AI adoption rather than blocking it. It creates predictable pathways from idea to mission-aligned, well-controlled deployment, reducing crisis-driven rework.

2 From IT Governance to AI Governance

2.1 Opening vignette: "We Already Have Governance... Don't We?"

The IT governance committee had been running smoothly for years. It managed project approvals, architecture standards, security exceptions, and major changes to production systems. The agenda was predictable: infrastructure refreshes, ERP upgrades, security patches, vendor renewals. When the first AI proposal appeared on the docket—a risk-scoring model for internal audits—the chair reassured the group, "We already have strong governance. We'll run it through the same process." The team treated the AI project like any other system enhancement, checked it for uptime, security, and integration impacts, and approved it.

Six months later, internal auditors raised a concern: the AI model was consistently flagging certain departments more often than others, even when their underlying metrics were similar. The departments started pushing back, questioning the fairness of the risk scoring and the lack of transparency around how the model worked. The governance committee realized that all their standard questions about change control and system reliability had not prepared them to assess fairness, explainability, or model drift. They had governance—but not governance that was fit for AI.

This chapter is about that gap. You likely have IT governance structures in place today, including committees, policies, workflows, and standards. They are not obsolete, but they are incomplete for AI. Your mission is not to throw them away, but to evolve them—deliberately and pragmatically—so they can handle the realities of AI systems. This evolution is how you move from "we already have governance" to "we have the right governance for AI."

2.2 Why this evolution matters now

For many organizations, AI is arriving faster than governance can adapt. Business units experiment with generative AI tools for content, analytics, and decision support. Vendors offer pre-packaged AI features embedded in platforms you already rely on. Cloud providers expose powerful AI services through APIs that developers can call with just a few lines of code. Against this backdrop, sticking strictly to traditional IT governance is like bringing a firewall mindset to a world of zero-trust networks—it helps, but it is not enough.

AI is not just another application. It changes who or what is making decisions, how those decisions are justified, and how they evolve. Traditional governance focuses on systems that behave deterministically and change infrequently, whereas AI systems learn from data, adapt, and sometimes behave in ways stakeholders did not anticipate. If you apply yesterday's governance patterns to today's AI, you risk blind spots in ethics, accountability, and risk management.

The pressure is also external. Boards, regulators, and the public now recognize AI's impact on fairness, privacy, and safety. They are asking whether governance keeps pace with innovation. As an IT manager, your credibility depends on your ability to show that you are not just controlling infrastructure and applications, but also the intelligent systems built on top of them. That credibility comes from being able to say: "Yes, we have IT governance—and we have AI governance integrated into it."

2.3 Traditional IT governance: what you already know

Before you extend governance, it helps to be explicit about what you already have. Traditional IT governance usually revolves around a few core elements: decision rights, demand and portfolio management, architecture and standards, risk and security management, and compliance and audit. These elements have served organizations well through waves of mainframes, client/server, web, and cloud.

- Decision rights clarify who approves which investments, changes, and exceptions. They define which issues are resolved by project managers, which go to IT leadership, and which require executive or board-level input.

- Demand and portfolio management ensure that technology investments align with strategy. They help prioritize projects, allocate resources, and avoid overloaded technology roadmaps.

- Architecture and standards provide technical coherence.
  They define preferred technologies, integration

patterns, and design constraints to reduce complexity and cost over time.

- Risk and security management focus on confidentiality, integrity, and availability. They address threats such as cyberattacks, data loss, and unauthorized access, using policies, controls, and monitoring.

- Compliance and audit verify that systems meet regulatory requirements and internal policies. They provide documentation, evidence, and assurance to regulators, auditors, and internal oversight bodies.

These components remain important in the AI era. AI systems still need to be secure, resilient, and compliant. They still need to fit within your architecture and align with strategy. But traditional IT governance alone does not ask the questions that are specific to AI—such as who is accountable for model behavior, how fairness is defined and monitored, or how automated decisions are explained to affected stakeholders.

2.4 What stays the same when AI enters

The good news is that you do not have to start from zero. Several governance fundamentals stay the same when AI enters the picture. Understanding what remains stable helps you avoid overreacting and creating unnecessary complexity. It also helps you reassure stakeholders that AI governance builds on familiar structures rather than creating an entirely new regime.

- Strategy alignment still matters.
  AI projects must connect clearly to organizational objectives, just like any other significant technology investment. "Because AI is trendy" is still not a strategy.

- Clear decision rights are still critical.
  Someone must own AI use cases, approve deployments, manage exceptions, and handle escalations. Ambiguous ownership creates the same confusion in AI as in any other technology domain.

- Risk management remains central.
  You are still responsible for identifying, assessing, and mitigating risks. The risk categories may expand, but the discipline of risk management remains applicable.

- Documentation and audit trails remain necessary.
  You still need to show what was built, who approved it, how it is monitored, and what changes were made over time. AI adds new types of evidence, such as model validation reports and bias assessments, but the principle stays the same.

- Cross-functional collaboration is still essential.
  Security, privacy, legal, compliance, and business units already work together in IT governance. With AI, that collaboration deepens, but it does not come from nowhere.

Recognizing these continuities allows you to frame AI governance as an evolution rather than a revolution. You will add new questions, artifacts, and processes, but they plug into an existing skeleton. As a manager, this mindset helps you position AI governance as a modernization of your current practices rather than a competing governance structure.

2.5 What changes: from static systems to learning systems

Where the differences emerge is like AI systems themselves. Traditional IT systems are largely static: once deployed, their behavior remains stable until a change is formally introduced through a release process. AI systems, particularly those built on machine learning and generative models, are fundamentally different. They learn from data, may be retrained, and can behave unpredictably, especially when exposed to new inputs or adversarial prompts.

This shift introduces new governance questions. Instead of just asking "What version of the software is in production?" you must also ask "What data was the model trained on, and when?" Instead of tracking only functional defects, you must also track performance drift, unfair outcomes, and unexpected behaviors. Instead of approving a single design, you must approve a lifecycle that covers data collection, training, evaluation, deployment, monitoring, and retirement.

AI also blurs boundaries between systems you control and those you consume. You may build some models internally, fine-tune others on vendor platforms, and consume many as black-box services. In each case, governance must define the level of assurance you require, the evidence you expect from vendors, and the residual risk you are willing to accept. This is fundamentally different from governing a server or a database that behaves in predictable ways and is fully under your configuration control.

2.6 Mapping IT governance domains to AI governance needs

To make this concrete, it helps to map your existing IT governance domains to new AI governance needs. Think of it as extending each domain with AI-specific questions rather than reinventing it.

- Project and portfolio governance must now consider AI risk tiers.
  A low-risk AI-powered search feature should not face the same scrutiny as an AI system that influences healthcare decisions or public benefits eligibility. You will need criteria to categorize AI initiatives by impact and risk.

- Architecture and standards must expand to include model types, data sources, and AI platforms.
  You will need to define approved AI tooling, integration patterns for models, and guidelines for where AI is appropriate (and where it is not).

- Risk management must include ethical, reputational, and systemic risks unique to AI.
  This means defining risk categories like bias, explainability gaps, and model misuse—not just traditional security incidents or outages.

- Security governance must extend to model and data protections.
  You must consider risks such as model theft, poisoning of training data, prompt injection, and leakage of sensitive information through AI interfaces.

- Compliance governance must account for emerging AI-specific regulations and internal policies.
  You may need new documentation forms, assessment templates, and attestation processes specific to AI systems.

By overlaying AI needs onto existing domains, you get a roadmap for where to adjust charters, processes, and templates. This is an incremental, manager-friendly way to evolve governance rather than standing up AI governance in isolation.

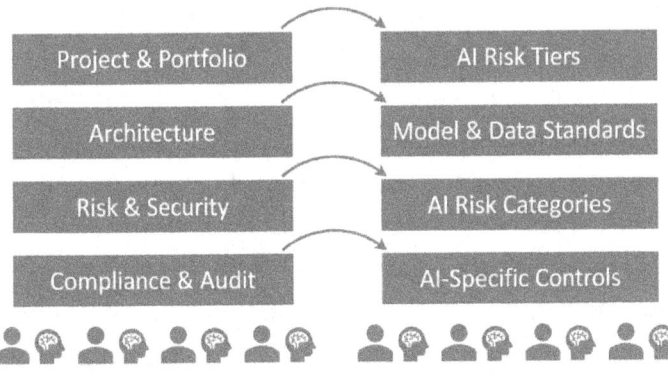

2.7 The AI operating model vs. the IT operating model

Most IT organizations already have an operating model: roles, processes, and structures that define how work gets done. It covers planning, delivery, support, and change. AI needs its own operating model, but it cannot be disconnected from IT. Instead, you need an AI operating model that sits within and extends your existing IT operating model.

An AI operating model clarifies questions such as: Who identifies AI opportunities? Who evaluates feasibility and risk? Who builds and validates models? Who approves deployment? Who monitors behavior and handles incidents? It defines these responsibilities across roles, including IT leadership, data science, security, compliance, and business units. Many organizations formalize this through structures such as an AI governance council, an AI center of excellence, or federated AI leads across business lines.

The relationship between IT and AI operating models is complementary. IT provides the underlying platforms, integration patterns, security controls, and lifecycle management processes. AI adds specialized capabilities: model development, data curation, experiment tracking, and bias and performance assessments. For governance, your task as a manager is to ensure that the AI operating model plugs into existing gates and workflows, rather than bypassing them. That way, AI initiatives follow consistent paths from idea to production, with added AI-specific checks where needed.

2.8 Roles and responsibilities: who leads AI governance?

One of the most common sources of friction in AI governance is unclear ownership. IT, data science, legal, compliance, risk, and business units all touch AI, but no single group can govern it alone. As AI becomes more pervasive, you need a deliberate allocation of responsibilities that reflects both technical and non-technical perspectives.

A practical approach is to think in layers: strategic, operational, and execution. At the strategic layer, senior leaders (CIO, CDO, Chief Risk Officer, and business executives) set direction, define risk appetite for AI, and approve high-level policies. At the operational layer, cross-functional governance groups define standards, approve significant AI use cases, and oversee monitoring and reporting. At the execution layer, delivery teams build, test, deploy, and operate AI systems in compliance with governance standards.

Your role as an IT manager often spans the operational and execution layers. You may sponsor AI initiatives, chair or participate in AI review boards, and ensure that AI systems follow architecture, security, and risk guidelines. You also act as a bridge between technical teams and non-technical stakeholders, translating governance requirements into actionable tasks and ensuring that AI projects do not bypass critical checks in the rush to deliver. Clarifying this role— and those of your peers—prevents AI governance from being everyone's responsibility in theory and no one's in practice.

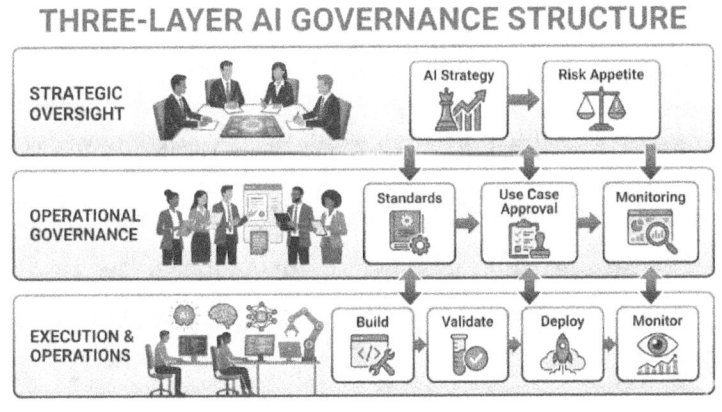

THREE-LAYER AI GOVERNANCE STRUCTURE

2.9 The manager's checklist: bridging IT and AI governance

To help you operationalize this evolution, use the following checklist when reviewing or designing governance processes for AI initiatives. Each item highlights a bridge between your existing IT governance and the AI-specific enhancements you need.

- Have we explicitly defined how AI projects enter our existing governance processes?
  This ensures AI is not handled informally or as "experiments" outside normal review channels.

- Do our project intake and approval templates include AI-specific fields (e.g., AI type, risk tier, affected stakeholders)?
  This helps you spot high-risk AI uses early and route them appropriately.

- Are architecture review boards prepared to assess data flows, model dependencies, and AI platforms—not just applications and infrastructure?
  This expands architecture governance to cover model hosting, data pipelines, and integration patterns for AI services.

- Do security and privacy reviews explicitly consider AI-related threats (e.g., data leakage via prompts, model poisoning, adversarial inputs)?
  This prevents AI-driven risks from falling through the cracks of traditional security reviews.

- Is there a defined path to escalate AI use cases that involve high-stakes decisions (e.g., eligibility, safety, employment) to senior oversight?
  This ensures that the most consequential AI initiatives receive appropriate scrutiny and leadership attention.

You can use this checklist as a simple gauge of AI-readiness for your existing governance. If the answer to most questions is "no" or "not yet," you have a clear agenda for modernization.

2.10 The AI Reality Check: Avoiding parallel governance universes

A common reaction to AI is to stand up a separate AI governance framework, complete with new committees, policies, and forms. While some dedicated structure is necessary, creating completely parallel governance universes—one for IT and one for AI—often backfires. It creates confusion about which rules apply, duplicates effort, and tempts teams to route work through the path of least resistance, undermining both regimes.

The reality is that AI governance must sit inside, and partially reshape, your broader IT governance and risk framework. You need shared principles, shared templates, and shared oversight structures. When AI is treated as a special case with a separate governance system, its decisions may not align with enterprise risk appetite, architectural constraints, or strategic priorities. Conversely, when AI is forced to operate under unchanged IT governance, key AI-specific risks are ignored.

The goal is disciplined convergence: one governance framework that recognizes AI as a distinct class of technology with unique risks and controls, yet still reports to the same leadership structures, risk committees, and audit processes. As a manager, you are in a key position to spot and avoid the trap of parallel universes. Your influence on process design, committee charters, and approval workflows can help keep AI governance integrated and coherent.

2.11 The manager's playbook: conversations to have this quarter

To translate this chapter into action, here are concrete conversations you can initiate over the next quarter. They do not require new budgets or tools—only leadership and focus.

- With your governance or change advisory board chair:
  Ask how AI projects are currently identified and handled. Explore adding explicit AI categories and risk tiers to the agenda and intake forms.

- With your architecture review lead:
  Discuss creating simple AI architecture patterns that cover model hosting, data flows, and integration with existing applications and APIs.

- With your security and privacy leads:
  Identify AI-specific threats relevant to your environment and agree on how they will be assessed

during reviews (e.g., prompt leakage, data residency in AI services).

- With your legal, compliance, or risk officers:
  Clarify emerging regulatory expectations and how to document AI use cases, decisions, and controls to satisfy auditors and oversight bodies.

- With your business or mission owners:
  Align on where AI is most appropriate and where it is not, and how governance will help them move faster and safely, rather than slowing them down.

Each of these conversations helps close the gap between IT governance as it exists today and AI governance as it needs to be. They also reinforce your role as a manager who brings structure and clarity to a rapidly changing landscape.

2.12 How this chapter sets up the next steps

In this chapter, you saw that AI governance is not a separate universe, but an evolution of IT governance focused on systems that learn and decide. You explored what stays the same, what changes, how to extend existing governance domains, and how to clarify roles and operating models. You also gained a practical checklist and conversation prompts to help you move your organization forward.

Next, in Chapter 3, we will go deeper into the **AI operating model** itself. We will define the core components of an AI operating model for IT managers, describe common patterns (centralized, federated, and hybrid), and show how to embed "responsible by design" practices into the way AI work gets done. You will move from conceptual mapping to concrete structures that can be implemented, piloted, and refined in your environment.

---

3 Building the AI Operating Model

3.1 Opening vignette: "The AI wins, the workflow loses."

The IT manager for a large healthcare network was proud of the new AI-powered triage assistant deployed in the contact center. The model accurately categorized patient calls, suggested next-best actions, and projected call volumes throughout the day. The pilot metrics looked strong: shorter handle times, higher first-call resolution, and fewer transfers. Leadership hailed it as a model modernization initiative and asked for a rollout across all locations.

Within weeks of expansion, problems began to surface. Nurses complained that the AI recommendations did not align with the actual complexity they observed upon patients' arrival. The scheduling team reported mismatches between projected and actual appointment types, causing overstaffing in some clinics and understaffing in others. The AI itself "worked," but the workflows around it had not changed. No one had redefined roles, adjusted processes, or clarified who could override AI recommendations and when.

The result was friction. Frontline staff felt that AI was being imposed on them without a clear operating model. Managers received conflicting feedback: "The metrics look great" versus "This creates chaos on the floor." Eventually, the CIO realized the core issue: they had introduced an AI **capability** without designing an AI **operating model**. This chapter helps you avoid that mistake by showing how to make AI part of the way your organization works, not just a set of disconnected tools.

3.2 What is an AI operating model (for managers)?

An AI operating model describes how your organization identifies, builds, deploys, and manages AI solutions in a repeatable, controlled, and mission-aligned way. It is the blueprint for "how we do AI here." It defines roles, responsibilities, processes, decision rights, and enabling platforms that support AI across its lifecycle—from idea to production and continuous improvement.

Unlike a one-off AI project plan, an operating model is durable. It survives staff turnover, vendor changes, and technology upgrades. It gives your teams a common language and a shared structure: where AI ideas go, who evaluates them, how risk is assessed, who approves deployments, and how performance is monitored. As an IT manager, you use the AI operating model as your lever to scale AI responsibly, rather than repeatedly reinventing governance for each new initiative.

At its core, an AI operating model should be **manager-ready**: understandable without deep data science expertise, actionable within existing IT structures, and aligned with your organization's value, risk, and control expectations. You do not need a perfect model from day one; you need a clear starting point that can evolve as your AI maturity grows.

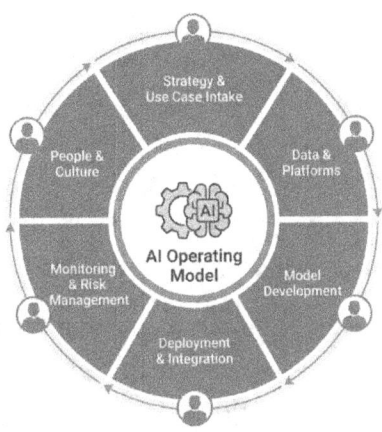

3.3 Core components of an AI operating model

Your AI operating model should be composed of a small number of well-defined components. Think of these as building blocks you can tailor to your organization.

- Strategy and use case intake
  This component defines how AI ideas are submitted, evaluated, and prioritized. It aligns AI initiatives to mission, value, and feasibility and prevents random experiments from bypassing governance.

- Data, platforms, and tools
  This covers the data foundations and technical

platforms that AI depends on: data governance, feature stores, model repositories, MLOps tools, and approved AI services. It gives teams a safe "railroad" to build on instead of ad hoc stacks.

- Model development and validation
  This defines how models are designed, trained, tested, and documented. It includes standards for bias assessment, performance metrics, and explainability, ensuring that AI solutions meet your governance expectations before deployment.

- Deployment, integration, and change management
  This addresses how AI is embedded into existing systems and workflows. It defines release processes, rollback mechanisms, and how AI updates are coordinated with business change management.

- Monitoring, incident management, and lifecycle governance
  This governs how AI performance, fairness, and risk are monitored over time. It defines thresholds, alerts, review cadences, and processes for decommissioning or retraining models.

- People, roles, and culture
  This component defines who does what: sponsorship, data science, engineering, operations, risk, and frontline users. It also covers training, communication, and incentives to support a governance-aware AI culture.

Taken together, these components form the skeleton of your AI operating model. The details will differ by organization, but the categories give you a checklist for completeness: if a component is missing or vague, that is where problems will eventually appear.

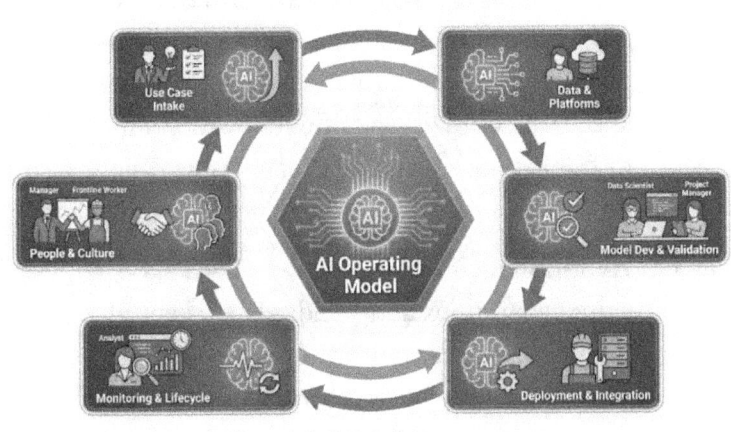

3.4 Centralized, federated, and hybrid AI operating models

One of the most important design choices is whether your AI operating model should be centralized or decentralized. There is no single correct answer; the right pattern depends on your size, complexity, risk profile, and current level of AI maturity. As a manager, you should understand the trade-offs of three common patterns.

- Centralized model
  A central AI team (often a center of excellence) owns most AI development, standards, and governance. Business units request support from

this team. This pattern offers strong consistency and control but can become a bottleneck if demand outstrips capacity.

- Federated model
  Business units have their own AI capabilities, with a light central function providing standards, platforms, and coordination. This pattern supports speed and domain-specific innovation but can lead to fragmentation and duplicate effort if not carefully governed.

- Hybrid model
  Core capabilities and standards are centralized (platforms, governance frameworks, shared services), while business units build and own AI use cases within those constraints. This pattern aims for both control and agility but requires clear role definitions and strong communication.

For many organizations, especially in the early to mid stages of AI maturity, a hybrid model is the most practical. It lets you build a central spine of governance and platforms while empowering business-aligned teams to innovate within defined guardrails.

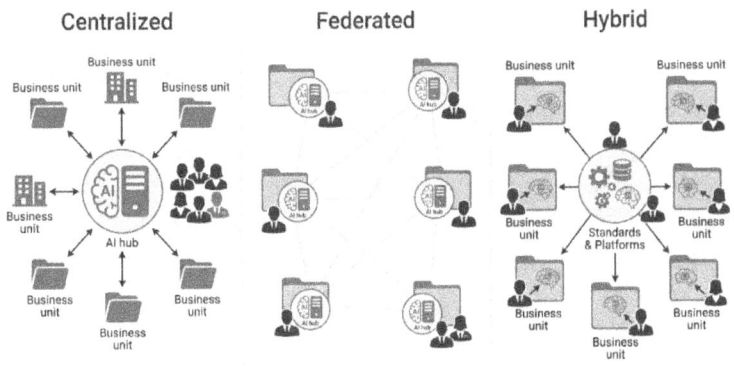

Centralized	Federated	Hybrid

3.5 Designing your AI operating model: a step-by-step approach

You do not need to design your AI operating model in a single giant effort. Instead, take a stepwise approach that can be executed over a few months and refined over time.

Step 1: Map your current state
Document how AI projects currently happen: who initiates them, who builds them, what platforms they use, how they get approved, and how they are monitored (if at all). The goal is to make the implicit explicit.

Step 2: Identify pain points and risks
From recent AI or advanced analytics initiatives, list where things went wrong or felt chaotic—such as unclear ownership, poor integration, no monitoring, or surprise regulatory questions. These pain points are your strongest arguments for a more structured operating model.

Step 3: Choose a pattern (centralized, federated, hybrid)
Based on your maturity and constraints, pick a starting pattern. You can evolve, but you need an initial stance so people know where decisions are made and who to call.

Step 4: Define key roles and decision rights
Clarify who approves AI use cases, who owns data, who builds models, who signs off on risk assessments, and who can halt or roll back a model. Write these decision rights down and socialize them.

Step 5: Align with existing IT governance
Integrate AI checkpoints into your existing governance gates (project intake, architecture review, security review, change management). Avoid creating standalone AI processes that bypass established structures.

Step 6: Pilot, learn, and refine
Apply the emerging operating model to a small number of AI initiatives. Observe where it works and where it adds unnecessary friction. Refine roles, workflows, and templates based on feedback.

This incremental approach reduces resistance and lets you adapt the model to your organization's realities rather than importing a theoretical blueprint.

Six Steps to an AI Operating Model

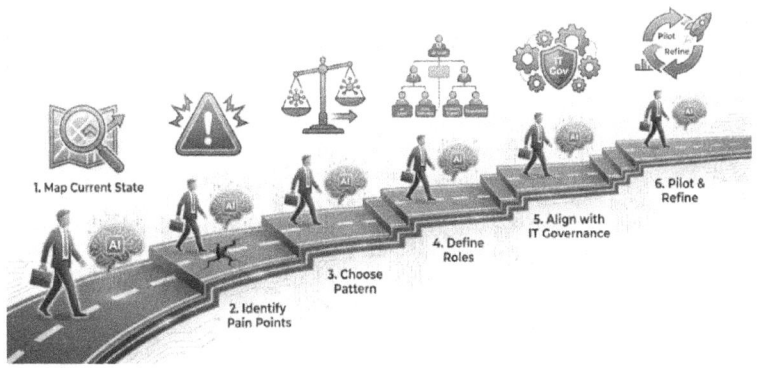

1. Map Current State
2. Identify Pain Points
3. Choose Pattern
4. Define Roles
5. Align with IT Governance
6. Pilot & Refine

3.6 Key roles in the AI operating model

Your AI operating model will only function if people understand their roles. While job titles differ, the underlying responsibilities are similar across organizations. As an IT manager, you should know which roles must exist, even if they are combined in a small team.

- AI sponsor (often a business or mission leader) Owns the business outcome, champions the AI initiative, and ensures alignment with mission and strategy. This role prevents AI from becoming "tech for tech's sake."

- Product or use case owner Defines the problem, success metrics, and constraints. Acts as the bridge between business, IT, and data science, ensuring the solution solves real workflow problems.

- Data owner and steward
  Manages data access, quality, and governance for AI datasets. Ensures compliance with privacy, security, and ethical standards related to data use.

- Data scientist or ML engineer
  Designs, trains, and evaluates models. Works within governance standards to document assumptions, performance, and limitations of AI systems.

- IT/platform engineer
  Integrates AI into production systems, ensures scalability, reliability, and security, and maintains the underlying infrastructure and tooling.

- Risk, compliance, and security leads
  Assess AI-specific risks, review documentation and testing evidence, and ensure alignment with regulatory and internal policy requirements.

- Frontline users and operational managers
  Interact with AI in their daily work, provide feedback on usability and impact, and exercise human oversight where required.

You may not have all these roles neatly separated, especially early on. The important thing is that the responsibilities are clearly assigned and visible to everyone involved in AI initiatives.

3.7 The "responsible by design" loop

A central idea of this series is being **responsible by design**. In the context of your AI operating model, this means embedding responsibility, safety, and governance into every stage of the AI lifecycle, rather than bolting them on at the end. The operating model should ensure that no AI system can proceed without passing key governance checkpoints.

In practical terms, this looks like: AI ideas are screened not just for value, but also for potential risk and stakeholder impact. Data is vetted for quality and fairness before model training begins. Model evaluation includes bias and explainability tests, not just accuracy. Deployment plans specify who will oversee AI behavior and how issues will be escalated. Monitoring dashboards track fairness, performance, and usage patterns, not only technical uptime.

The "responsible by design" loop is continuous. Feedback from operations, incidents, and changing environments flows back into model retraining, policy updates, and operating model refinements. The loop makes your AI governance living and adaptive, rather than a one-time design exercise.

Responsible by Design Loop

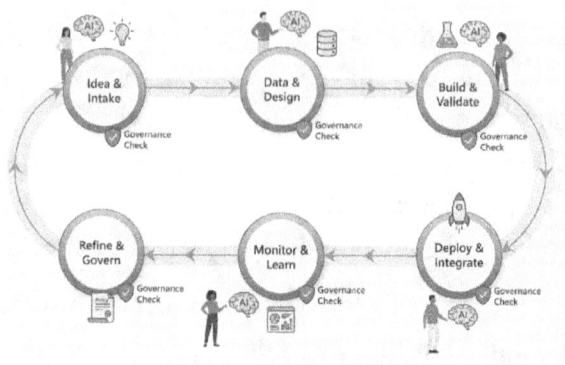

3.8 Practical example: applying the operating model to a use case

Consider a public-sector agency that wants to use AI to prioritize inspections for safety compliance. Without an operating model, the initiative might proceed informally: a team gets data, builds a model, and rolls it out as a dashboard. With an AI operating model, the process becomes structured and governed.

Use case intake ensures that the AI idea is documented, with clear objectives (e.g., reduce inspection backlog while maintaining fairness), affected stakeholders, and risk tier. Data and platform components ensure that only approved datasets are used, with documented lineage and quality checks. Model development follows standard templates: defining performance metrics, fairness criteria (e.g., no unfair disadvantage to specific neighborhoods), and explainability requirements.

Deployment and integration align with existing inspection workflows: how inspectors receive AI recommendations, when they can override them, and how overrides are logged. Monitoring tracks not only model accuracy but also inspection outcomes and potential disparities across regions or demographics. People and cultural elements ensure inspectors are trained to interpret AI suggestions and understand their own responsibilities.

This example shows that the AI operating model is not theoretical overhead; it directly shapes how AI is built and used, reducing the risk of misalignment with mission and stakeholder expectations.

3.9 The Manager's checklist: stress-testing your AI operating model

To test whether your AI operating model is more than a slide deck, ask yourself these questions:

- If a new AI idea appears tomorrow, do we know exactly where it goes and who reviews it? If the answer is unclear, your intake process and roles need refinement.

- For our existing AI systems, can we quickly show who owns each one, which data it uses, and how it was   validated? If this information is scattered or missing, your lifecycle documentation is weak and may fail under scrutiny.

- Do frontline staff know how to escalate concerns about AI behavior and feel empowered to do so? If not, your operating model may look good on paper but is failing at the human interaction layer.

- Are monitoring dashboards actively used, or do they exist but rarely drive decisions? If dashboards are not connected to clear thresholds and actions, monitoring is not truly integrated into your operating model.

- When regulations or internal policies change, do we have a mechanism to update AI standards and propagate them to teams? If policy updates rely on ad hoc emails and meetings, the operating model lacks resilience.

These questions help you identify where to strengthen the model, especially at the intersections between process, technology, and people.

3.10　How this chapter connects forward

In this chapter, you translated the abstract idea of AI governance into a concrete AI operating model that defines "how we do AI here." You examined its core components, common structural patterns, key roles, and the importance of being responsible by design. You also saw how to build your operating model incrementally and stress-test it with practical questions.

The next chapters dive into specific layers of that operating model. Chapter 4 focuses on the **governance framework for AI systems**—the policies, committees, and decision rights that sit atop the operating model and give it authority. Chapter 5 then zooms in on the data, showing how data governance and AI governance interlock. Together, they will give you the tools to refine your operating model into a robust, mission-aligned engine for AI modernization.

4 The Governance Framework for AI Systems

4.1 Opening vignette: "Who actually approved this?"

The quarterly risk review meeting had an unusual item on the agenda: a complaint from a major stakeholder about an AI-powered eligibility tool used to determine access to a social program. The tool had quietly gone live three months earlier, integrated into an existing web portal. At first, it seemed to work well—processing applications faster and reducing manual review queues. But recently, advocacy groups began to raise concerns that certain communities appeared to be denied at higher rates.

As the CIO, risk officer, legal counsel, and program director assembled, a simple question exposed a deeper problem: "Who actually approved this AI system?" The program team thought IT had approved it because it passed security checks. IT assumed the program team owned the business rules. Legal believed they had already signed off, but it turned out they had only reviewed a standard privacy notice. No one had holistically reviewed the AI system's purpose, risk, impacts, or fairness. There were pockets of governance—but no integrated **governance framework for AI systems**.

This chapter answers the question that meetings could not: who approves, on what basis, and under what conditions. It shows you how to design and implement an AI governance framework that is clear, repeatable, and aligned with your existing IT and risk structures—so you never face that question unprepared again.

4.2 What is an AI governance framework (in practical terms)?

An AI governance framework is the structured set of principles, policies, roles, and processes that determine how AI systems are proposed, evaluated, approved, monitored, and retired. It is the formal expression of "how we make decisions about AI," not just technically but organizationally. For managers, it is the bridge between strategy and day-to-day AI activity.

A good framework does three things. First, it clarifies **decision rights**: who can approve what, and when escalation is required. Second, it defines **processes and artifacts**: the steps AI systems must follow (e.g., risk assessment, testing, documentation) and the evidence they must produce. Third, it establishes **oversight mechanisms**: committees, review boards, and reporting that keep leadership informed and empower them to intervene when needed.

Most importantly, the framework should be **fit for purpose**. It must be strong enough to handle high-risk AI scenarios, but flexible enough not to strangle low-risk, low-impact experimentation. It is not about controlling everything. It is about applying the right level of governance to the right AI system at the right time.

AI Governance Framework Overview

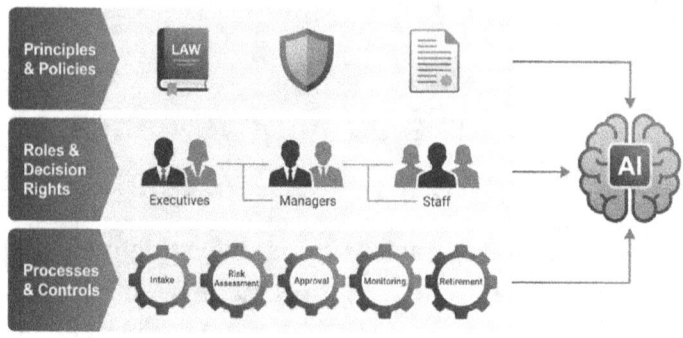

4.3 Principles: the foundation of your framework

Before you define committees or workflows, you need clear principles. These principles are short, explicit statements that guide all AI decisions. They are the lens through which proposals are evaluated, and trade-offs are made. Without principles, your governance becomes a checklist exercise; with them, it gains direction and coherence.

Typical AI governance principles might include:

- Mission alignment
  AI systems must support clearly defined mission or business outcomes and must not undermine the organization's core purpose or public trust.

- Proportionality of control
  The level of governance, testing, and oversight must be proportional to the AI system's potential impact

and risk. High impact and high risk require stronger controls.

- Human accountability
  AI may assist in decisions, but humans remain accountable for outcomes. Roles and responsibilities must explicitly reflect this.

- Transparency and explainability
  AI decisions should be explainable at a level appropriate to stakeholders, enabling meaningful oversight and recourse.

- Fairness and non-discrimination
  AI systems should be designed and monitored to avoid unfair bias and disproportionate negative impact on particular groups.

- Safety and robustness
  AI systems must be resilient to errors, misuse, and relevant cyber threats, with clear fallbacks when failures occur.

These principles become the "north star" of your governance framework. They should be simple enough to remember, visible in policies and templates, and referenced in review discussions.

4.4 Policies and standards: codifying expectations

Principles become real when they are translated into policies and standards that people must follow. For AI, these policies should extend, not duplicate, your existing IT, data, security, and risk policies. They answer questions like: What documentation is required for AI systems? What testing must be done before deployment? What monitoring is mandatory? Who can approve exceptions?

Examples of AI-related policies and standards include:

- An AI Use Policy
  Defines what types of AI use are allowed, restricted, or prohibited (e.g., no AI for fully automated hiring decisions without human review).

- An AI Risk Assessment Standard
  Specifies the dimensions to evaluate (e.g., impact, bias, explainability, security, privacy) and the risk tiers (low, medium, high, critical).

- An AI Model Documentation Standard
  Requires a consistent "model card" or similar artifact describing purpose, data sources, assumptions, performance metrics, limitations, and intended users.

- An AI Monitoring and Incident Policy
  Defines what must be monitored, how anomalies are detected, how incidents are classified, and how they are reported and resolved.

- An AI Vendor and Third-Party Policy
  Describes requirements for vendors providing AI capabilities, including transparency expectations, evidence of testing, and contract clauses.

You do not need dozens of policies. Start with a small set that covers the main lifecycle phases and risk drivers, then expand if needed. The key is that these policies are integrated into your broader governance documentation, not floating as disconnected PDFs.

Diagram 4.2: **From Principles to Policies to Practice**

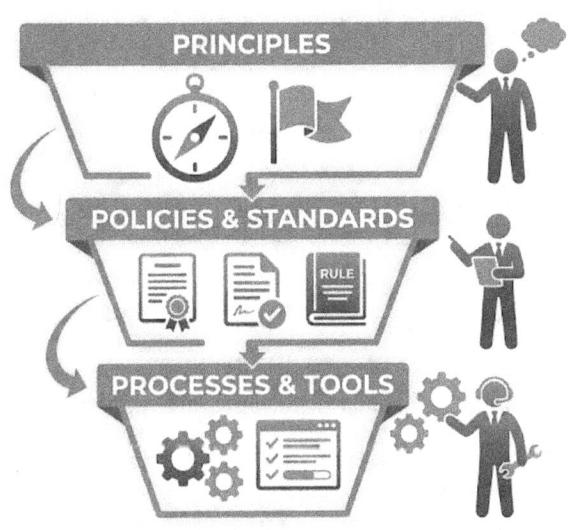

4.5 Roles and decision rights: who decides what

At the heart of the governance framework are decision rights: clear definitions of who can make which decisions about AI. Without this clarity, you end up with AI projects launched under the radar or stalled because approvals are unclear. For managers, knowing where your role fits in this structure is essential.

You can structure AI decision rights across four levels:

- Strategic level
  Approves AI strategy, risk appetite, and enterprise-level principles. Usually, the board or executive leadership is often advised by a senior risk committee.

- Portfolio level
  Approves or rejects high-impact AI use cases, assigns ownership, and ensures alignment with strategy and risk appetite. Often, a cross-functional AI or technology governance council.

- Project level
  Oversees design, development, and testing of specific AI systems; ensures compliance with standards before deployment. Typically includes IT, data science, business sponsors, and risk or compliance representatives.

- Operational level
  Makes day-to-day decisions about AI operation: model retraining, parameter tuning, incident response, and minor changes within established boundaries. Often handled by product owners, operations managers, and SRE/MLOps teams.

For each level, you should explicitly define: what decisions are made, what criteria must be met, what documentation is required, and when escalation to a higher level is necessary. This prevents both under-governance (no one truly approves) and over-governance (everyone must approve everything).

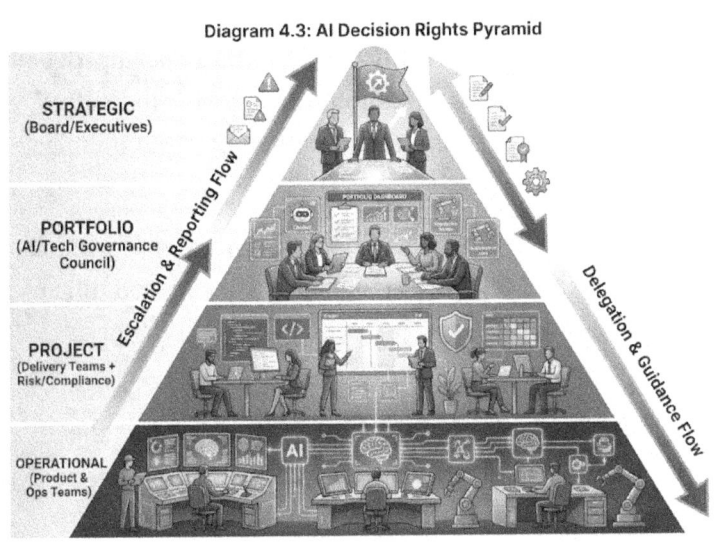

Diagram 4.3: AI Decision Rights Pyramid

4.6 Committees and councils: structuring oversight

Most organizations use committees or councils to coordinate governance. For AI, the goal is not to create a new bureaucracy, but to ensure that AI considerations are systematically discussed at the right forums. You can either integrate AI into existing committees or create specialized structures where needed.

Common patterns include:

- AI or Digital Ethics Council
  Focuses on the ethical, societal, and fairness

implications of AI use cases, especially high-stakes ones. Provides guidance and, in some cases, veto power.

- AI Governance Council or Board
  Oversees the portfolio of AI initiatives, approves high-risk use cases, monitors key risk indicators, and ensures alignment with strategy and policies.

- AI Technical Review Board
  Reviews model design, data choices, performance tests, and technical risks. Often composed of senior data scientists, architects, and security experts.

- Integration into existing IT governance bodies
  For example, adding AI-specific agenda items to change advisory boards, architecture review boards, and security councils, rather than duplicating structures.

The exact configuration will vary, but the principle is constant: AI must have a home in your governance forums. AI topics should not appear only as one-off agenda items during incidents; they should be part of recurring oversight and planning.

4.7 Processes: the lifecycle of AI decisions

Your governance framework comes to life through processes that define the AI lifecycle: from ideation to retirement. These processes should align with the AI operating model you defined in the previous chapter, but now with explicit governance steps.

A typical governance-aware AI lifecycle process includes:

- Intake and categorization
  AI ideas are submitted through a standard channel, documented, and categorized by risk tier and domain. Low-risk ideas may follow a lighter path; high-risk ones trigger deeper review.

- Preliminary assessment
  A quick evaluation checks strategic alignment, potential impact, data availability, and obvious red flags (e.g., legal constraints, prohibited uses).

- Detailed risk and impact assessment
  For non-trivial AI systems, teams complete a structured assessment covering fairness, explainability, security, privacy, safety, and stakeholder impact.

- Design and testing review
  Models and systems are reviewed for compliance with standards before deployment. This includes reviewing test results, documentation, and mitigation plans.

- Deployment approval
  The appropriate authority (based on risk tier) approves the system for production with defined scope, conditions, and monitoring expectations.

- Ongoing monitoring and periodical review
  AI systems are monitored for performance and risk, with periodic reviews to determine whether to continue, retrain, modify, or retire.

By defining these processes clearly and making them visible, you turn governance from an ad hoc conversation into a predictable pathway that teams can plan around.

4.8 The AI risk tiering model: focusing governance where it matters most

Not all AI systems are equal in their potential impact. A simple AI feature that suggests search terms is not in the same risk category as an AI system that helps decide who receives healthcare, financial services, or public benefits. A

risk tiering model allows you to apply governance proportionally, focusing effort where it matters most.

A practical tiering model might classify AI systems into four levels:

- Tier 1 – Minimal impact
  AI systems have a limited impact on individuals and no material effect on rights, access, or safety. Governance is light: basic documentation and monitoring.

- Tier 2 – Moderate impact
  AI systems are influencing decisions, but with human oversight and low potential for serious harm. Governance includes structured risk assessment and standard approval.

- Tier 3 – High impact
  AI systems that significantly affect access to services, resources, or opportunities, or that operate at scale. Governance includes detailed risk assessments, fairness and explainability testing, and higher-level approval.

- Tier 4 – Critical impact
  AI systems affecting safety, fundamental rights, or highly sensitive areas. Governance includes the strictest controls: executive-level approval, ongoing external scrutiny where appropriate, and robust incident response plans.

Each tier should have clearly defined governance requirements: which templates to use, which forums must review, what evidence is needed, and who has final approval rights.

AI Risk Tiers and Governance Intensity

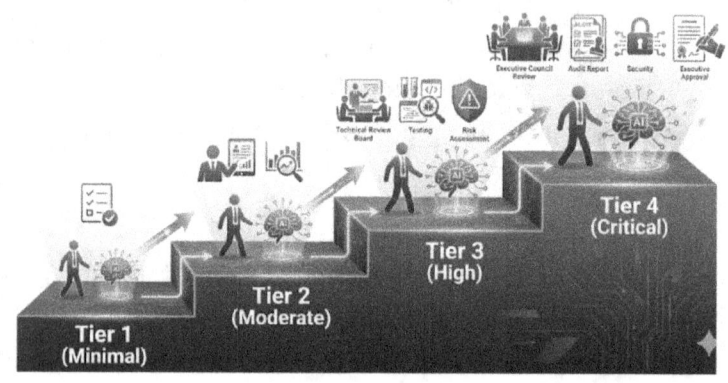

4.9 The manager's checklist: testing your AI governance framework

To see whether your governance framework is robust and usable, challenge it with these questions:

- Can we name the principles guiding our AI decisions, and do teams recognize them?
  If principles exist only in a slide deck, they are not yet part of your framework in practice.

- Do we have a clear, documented process for approving AI systems by risk tier?
  If approvals are ad hoc or personality-driven, your framework needs more structure.

- Is there a single place where we can see all AI systems in production, their tier, owner, and last review date?
  If not, your governance lacks visibility, which is a major risk in itself.

- Do AI vendors go through a defined evaluation process that addresses transparency, risk, and compliance expectations?
  If vendor AI is treated like a black box, your governance framework has a major blind spot.

- When an AI incident occurs (e.g., suspected bias or failure), is there a documented incident response path?
  If response depends on who happens to notice and who they call, your framework is not yet operational.

If you find gaps, treat them as a roadmap. Each "no" or "unsure" is an opportunity to improve the framework before an incident forces the issue.

4.10 AI Reality Check: avoiding "checkbox governance."

A frequent failure mode is "checkbox governance": forms are filled, committees meet, and standards exist—but no one truly challenges assumptions or understands the AI systems in depth. In checkbox governance, the goal becomes passing the process, not ensuring responsible, mission-aligned behavior. This is particularly tempting when organizations are under pressure to both move fast and show that they are "doing something" about AI risk.

To avoid this, your framework must be anchored in **real questions and decisions**. Templates should prompt meaningful reflection, not just boilerplate text. Committees should include people empowered to say "no" or "not yet," and they should have enough time and expertise to examine critical use cases. Metrics should track not just process completion (e.g., "risk assessment done") but also outcomes (e.g., "number of AI incidents," "fairness measures over time").

As a manager, you set the tone. If you treat governance as a strategic tool to reduce risk and build trust, your teams will follow. If you treat it as a compliance exercise to be minimized, they will focus on getting past it with as little effort as possible. The framework is only as strong as the culture and leadership behind it.

4.11 How this chapter links to practice and the next chapters

In this chapter, you translated the idea of "governing AI" into a concrete governance framework: principles, policies, roles, decision rights, processes, and risk tiers. You saw how to structure oversight through councils and boards, integrate AI into existing governance forums, and avoid both under-governance and checkbox governance. You now have a mental model and a language for designing or refining your organization's AI governance framework.

In the next chapters, we will zoom in on key components of this framework. Chapter 5 focuses on **Data Integrity and Accountability**, showing how data governance and AI governance reinforce each other. Chapter 6 explores **Risk, Compliance, and Responsible AI**, providing more detailed tools for risk assessment, control selection, and documentation. Together, they will help you apply your governance framework to the two most critical aspects of AI systems: the data they rely on and the risks they create.

5 Data Integrity and Accountability

5.1 Opening vignette: "The model wasn't wrong, the data was."

The fraud analytics team at a large financial services organization proudly showcased its new AI model to senior leadership. The model flagged suspicious transactions with unprecedented accuracy—at least according to the testing dashboard. After deployment, alerts spiked, investigations increased, and leadership initially interpreted this as a sign that the model was uncovering previously hidden fraud. But within weeks, investigators complained that most alerts led nowhere. They were drowning in false positives, and legitimate customers were having their accounts frozen or challenged.

When the IT manager and data science lead dug into the problem, they discovered that the model had been trained on historical data that was never fully cleansed. Old system glitches, misclassified cases, and inconsistent labeling had all been absorbed as "truth." The model was faithfully amplifying the noise embedded in years of poor data hygiene. Technically, the model worked as designed—but the data did not. The real governance failure was not in AI governance alone; it was in **data integrity and accountability**. This chapter is about preventing that failure in your organization.

5.2 Why data integrity is central to AI governance

AI systems are only as reliable, fair, and trustworthy as the data they consume. If your data is incomplete, inaccurate, biased, or poorly governed, no amount of algorithmic sophistication can compensate. For managers, this means that AI governance and data governance cannot be separated. You cannot govern AI responsibly if you treat data as an afterthought or as "someone else's problem."

Data integrity goes beyond basic data quality. It includes accuracy, completeness, consistency, timeliness, lineage, and protection from unauthorized changes. In AI, weaknesses in any of these dimensions can lead to flawed predictions, unfair outcomes, and incorrect decisions at scale. Even more critically, if you cannot trace how data flows into and through your models, you cannot explain or defend AI decisions when stakeholders, auditors, or regulators ask hard questions.

In practical terms, your AI governance framework must sit atop a robust data governance foundation. You must know where training, validation, and operational data come from, how they are prepared, and who is accountable for their condition. Otherwise, you are building powerful AI capabilities on an unreliable substrate—an invisible risk that will eventually surface in very visible ways.

Data Foundation Pyramid

Enter executive presentation slide

Decisions & Outcomes

AI Models & Systems

Data Governance & Integrity

5.3 Defining data integrity for AI systems

Data integrity for AI is about ensuring that data remains accurate, consistent, complete, and trustworthy throughout its lifecycle—from collection and storage to preparation and modeling, and through production use and archival. For AI, this definition needs to be specific and actionable; high-level slogans like "good data" are not enough.

Key dimensions of data integrity in AI include:

- Accuracy
  Data should correctly represent the real-world entities or events it describes. Inaccurate labels or measurements directly distort model learning and outputs.

- Completeness
  Relevant fields should be populated and not

systematically missing for particular groups or scenarios. Systematic gaps can create hidden bias.

- Consistency
  Data should conform to consistent definitions, formats, and business rules across sources and time. Inconsistent coding (e.g., different department codes for the same unit) confuses models and complicates monitoring.

- Timeliness
  Data should be up to date for its intended use. Stale data can cause models to respond to realities that no longer exist, degrading performance and fairness.

- Lineage and provenance
  The origin, transformations, and usage of data must be traceable. Without lineage, you cannot understand how data quality issues arose or how to remediate them.

- Protection against unauthorized changes
  Access controls, audit trails, and change management must prevent and detect improper alterations to data that feed AI systems.

For managers, the goal is to embed these dimensions into data governance policies, tools, and practices, and then explicitly link them to AI initiatives.

5.4 Data governance and AI governance: a complementary pair

Data governance is not new. Many organizations already have structures to manage data quality, privacy, security, and access. The challenge in the AI era is to better **connect** those data governance practices to AI governance. Instead of creating separate, siloed efforts, you should view them as a complementary pair.

Data governance focuses on:

- Defining data ownership and stewardship
  Clarifying who is responsible for particular data domains and how they ensure quality and security.

- Setting data standards
  Establishing common definitions, formats, quality thresholds, and metadata requirements.

- Managing data access and protection
  Enforcing policies for who can access data, under what conditions, and for which purposes.

- Ensuring compliance with privacy and data regulations
  Handling consent, retention, residency, and usage restrictions.

AI governance builds on that foundation by:

- Defining acceptable AI uses of data
  Ensuring data is used in ways that respect context, consent, and fairness.

- Evaluating data suitability for training and inference
  Checking whether data is representative, sufficiently complete, and appropriate for the intended AI purpose.

- Requiring documentation of data used for models
  Capturing data sources, preparation steps, and limitations as part of the AI system documentation.

- Monitoring AI performance and outcomes back to data quality
  Connecting observed issues (e.g., bias, drift) to underlying data problems and governance gaps.

Seen this way, AI governance is the "consumer" that tests and validates whether your data governance is effective. Weak data governance will quickly manifest as AI failures. Strong data governance becomes a strategic advantage for reliable, responsible AI.

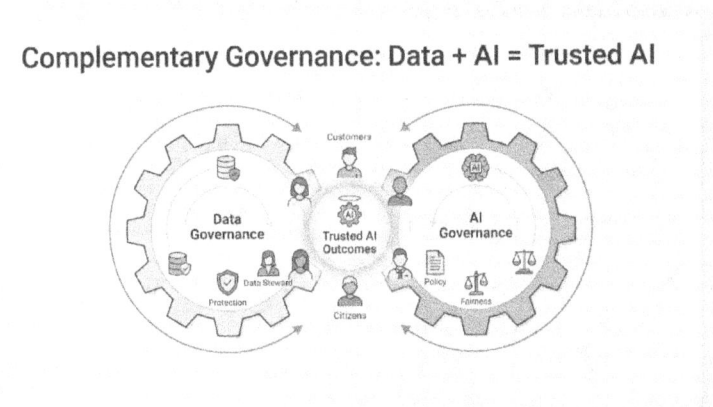

Complementary Governance: Data + AI = Trusted AI

5.5 Data accountability: who owns what in the AI lifecycle

Data integrity does not happen by accident; it happens because specific people are accountable for specific datasets and decisions. In AI, confusion often arises when everyone assumes someone else has checked the data. Your job as a manager is to make data accountability explicit.

Key data-related roles in AI systems include:

- Data owner
  Typically, a senior leader is responsible for a data domain (e.g., customer data, clinical data). Owns policies for data use, quality expectations, and access permissions.

- Data steward
  Operational custodian of data quality and metadata. Ensures that data definitions are clear, quality rules are implemented, and issues are addressed.

- Data engineer
  Designs and maintains pipelines that move data from source systems into analytics and AI platforms. Responsible for implementing quality checks and ensuring reproducibility.

- Data protection/privacy officer (or equivalent)
  Oversees compliance with data protection laws and internal privacy policies, ensuring that data used in AI adheres to consent and usage boundaries.

- AI product or model owner
  Accountable for how data is used in a specific AI

system, including training, validation, and monitoring. Coordinates with data owners and stewards.

To avoid gaps, you should define and document how these roles interact for each significant AI use case. When an AI initiative is proposed, data ownership and stewardship should be identified alongside the initiative's value and risks.

5.6 The AI data lifecycle: from raw to responsible

For governance, it helps to think in terms of an **AI data lifecycle**—a series of stages data passes through on its way to becoming input for AI models and decisions. At each stage, different integrity and accountability checks are needed.

Typical stages include:

- Collection and ingestion
  Data is collected from internal systems, external sources, or users, and ingested into data platforms. Governance concerns include consent, purpose limitation, and secure transfer.

- Storage and cataloging
  Data is stored in databases, data lakes, or warehouses and cataloged. Governance focuses on access control, classification, metadata, and retention policies.

- Preparation and feature engineering
  Data is cleaned, transformed, and combined to create training features. Governance checks include quality validation, handling of missing values, and bias analysis in feature distributions.

- Training and validation
  Data is used to train models and evaluate performance. Governance questions focus on representativeness, train/test splits, and consistency of performance across relevant subgroups.

- Deployment and operational use
  Data is fed into models in production. Governance includes monitoring data drift, ongoing quality checks, and ensuring that real-time data aligns with assumptions made during training.

- Archival and decommissioning
  Data used for historical models, debugging, and audits is retained or deleted in accordance with policy. Governance ensures you can reconstruct past decisions when needed, without storing more or longer than allowed.

At each stage, you should define what evidence is required—logs, reports, dashboards—and who reviews it. This lifecycle perspective prevents governance from being a one-time event at model training; it becomes continuous.

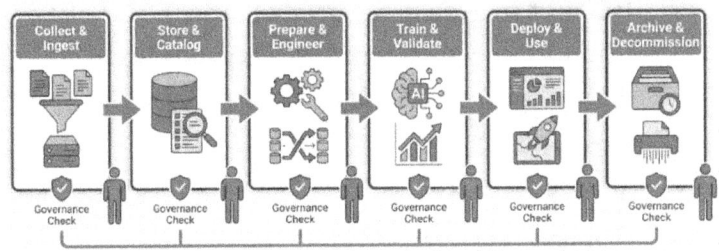

5.7 Data quality controls for AI: moving beyond generic checks

Traditional data quality programs often focus on generic metrics like completeness and format validation. AI requires a more nuanced approach because models are sensitive not just to individual record errors but to patterns and distortions across entire datasets. As an IT manager, you do not need to design every check, but you must ensure the **right types** of controls are in place.

Examples of AI-specific data quality controls include:

- Distribution checks
  Ensuring that key variables have realistic distributions and are not skewed in ways that misrepresent the population (e.g., only a small subset of groups appear).

- Label validation
  Verifying that target variables (e.g., "fraud,"

79

"approved," "diseased") are reliably and consistently labeled across time and sources.

- Bias and representativeness analysis
Checking whether the dataset composition disadvantages particular groups or scenarios, especially for protected characteristics or high-stakes decisions.

- Temporal consistency
Ensuring that the time dimension is handled properly (e.g., no future information leaks into past data) and that past data is still relevant for current and future decisions.

- Outlier and anomaly handling
Detecting unusual patterns that might indicate data errors, system integration problems, or malicious manipulation.

These controls should be standardized as part of your AI development and data preparation toolkits, not left to each project to invent from scratch. Governance can require that certain checks be performed and documented for given risk tiers.

5.8 Data lineage, observability, and auditability

When something goes wrong with an AI system—an unexpected decision, a biased report, a performance drop—the first question is often "what changed?" Without strong data lineage and observability, answering that question can be slow or impossible. From a governance standpoint, this is unacceptable; accountability depends on being able to trace data flows and transformations.

Data lineage tracks where data came from, which systems it passed through, what transformations were applied, and where it ended up. For AI, lineage must extend through feature engineering and model training: which datasets, at which versions, were used to train which model version. Data observability adds monitoring of data pipelines and quality in near real time: detecting anomalies, pipeline failures, and drift in data distributions.

Auditability means you can reconstruct how data contributed to AI decisions at a point in time. This might include logs of model inputs and outputs (subject to privacy rules), configuration of feature pipelines, and snapshots of model versions. When regulators, auditors, or affected individuals demand explanations or redress, you must be able to produce evidence. Without proper lineage, observability, and auditability, your AI governance is blind.

Diagram 5.4: Data Lineage and Observability for AI

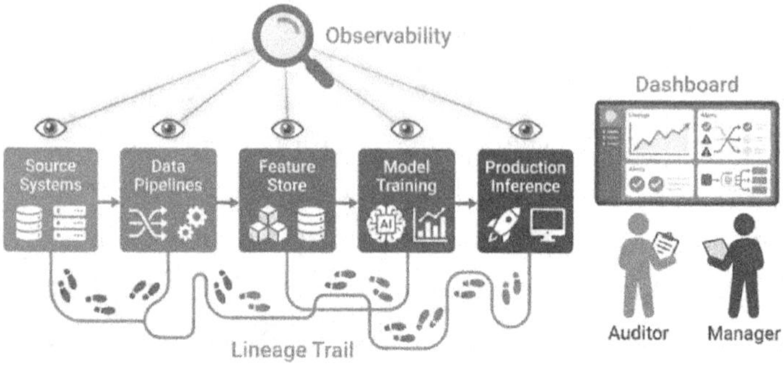

5.9 The manager's checklist: data integrity for AI projects

To operationalize data integrity and accountability in your AI initiatives, use this checklist as part of your project intake and review:

- Have we identified data owners and stewards for all major datasets used by this AI?
  If no clear owner exists, data quality and usage responsibilities are likely to fall through the cracks.

- Do we have documented data quality and integrity criteria specific to this AI use case?
  Generic "good enough" criteria are insufficient for high-impact AI; you need use-case-specific expectations.

- Has the dataset been analyzed for representativeness and potential bias across key groups?

If this analysis has not been done, you risk embedding historical inequities into AI behavior.

- Can we trace data lineage from source systems through transformation to the features used for training?
  If lineage is opaque, you will struggle to debug issues or answer accountability questions later.

- Is there a plan and tooling for monitoring data drift and quality in production?
  Without ongoing monitoring, gradual data degradation will silently undermine AI performance and trust.

You can embed these questions into standardized forms, templates, or governance tooling. Over time, they become part of your organization's reflexes whenever AI and data intersect.

5.10 AI Reality Check: You can't "fix it in the model."

In many organizations, there is a temptation to treat data problems as something data scientists can "fix in the model"—through clever algorithms, reweighting, or post-hoc corrections. While some issues can be mitigated at the model level, this mentality is dangerous. It shifts focus away from systemic data issues and places unrealistic expectations on modeling to compensate for poor data governance.

A healthier stance recognizes that some problems must be resolved at the **data governance** level. For example, if certain groups are underrepresented in your historical data because they were historically underserved, you cannot fully correct that bias with a few modeling tricks. You may need to change how data is collected, engage with stakeholders, or rethink the use case. Similarly, if your underlying transaction system has inconsistent coding practices, your long-term fix is to address that inconsistency at its source, not to patch it downstream continuously.

As a manager, your role is to push back against the "model will solve it" mindset and ensure that your AI governance framework always asks: "Is this a data problem, a model problem, or both?" This is how you build durable integrity and accountability rather than fragile workarounds.

5.11 How this chapter connects to the broader governance picture

In this chapter, you firmly brought data to the center of AI governance. You saw that data integrity and accountability are non-negotiable foundations for reliable and fair AI systems. You explored how data governance and AI governance reinforce each other, how to define data integrity for AI, and how to structure roles, lifecycle stages, controls, and observability. You also gained a practical checklist to help you apply these concepts to your projects.

In the next chapter, **Chapter 6 – Risk, Compliance, and Responsible AI**, we will build on this foundation to examine AI-specific risk categories, regulatory expectations, and practical mechanisms to demonstrate compliance and responsibility. You will learn how to integrate AI risk into your enterprise risk management, select appropriate controls by risk tier, and document AI systems in ways that satisfy both internal leadership and external scrutiny.

6 Risk, Compliance, and Responsible AI

6.1 Opening vignette: "The pilot that triggered an investigation."

A large agency launched an AI pilot to help prioritize case reviews. The idea seemed straightforward: use historical data to predict which cases were most likely to require escalation and route those to senior analysts first. The pilot went live under a "limited test" banner, and initial productivity metrics looked good—analysts cleared more cases per week and reported that the AI's recommendations were often "reasonable."

Two months later, an investigative journalist published an analysis suggesting that the AI was systematically rating certain communities as "lower priority," even when their case histories were similar to others. Advocacy groups quickly amplified the findings, and oversight bodies requested documentation: What risks had been identified? Who had approved the pilot? What monitoring was in place? The agency discovered that although security, privacy, and basic testing had been completed, there was no explicit AI risk assessment, no clear mapping to compliance expectations, and no documented rationale for the decisions the AI influenced. The pilot had unintentionally crossed a line from innovation into governance exposure.

This chapter is about ensuring you never end up in that position. It helps you understand AI-specific risk categories, align them with compliance obligations, and build a **responsible, proactive AI approach rather than a reactive one**.

6.2 Why AI risk feels different (and why that matters)

Traditional IT risk focuses on availability, security, and integrity of systems and data—questions like "Will the system stay up?", "Can attackers get in?", or "Will we lose data?" AI introduces new classes of risk that are harder to see and sometimes harder to quantify. These include unfair outcomes, opaque decisions, model misuse, and erosion of stakeholder trust.

The "feel" of AI risk is different because AI systems often operate in gray zones. They may not be clearly right or wrong, but they can be systematically skewed, misaligned with policy intent, or vulnerable to subtle manipulation. Their behavior may change over time as the environment changes or as models are retrained. The people affected may not even be aware that an AI system is involved, making harms slower to surface.

For you as a manager, this means AI risk cannot be treated as a simple extension of existing IT risk checklists. You need a structured approach to identify and categorize AI-specific risks, integrate them into your risk management processes, and ensure they are treated at the same level as cybersecurity, financial, or operational risks.

6.3 The AI risk landscape: key categories for managers

To make AI risk manageable, group it into categories you can systematically assess. A manager-ready AI risk lens typically includes:

- Ethical and fairness risk
  The risk that AI systems produce unfair or discriminatory outcomes, even unintentionally, for particular groups or individuals.

- Explainability and transparency risk
  The risk is that AI decisions cannot be meaningfully explained to stakeholders, regulators, or affected individuals, undermining accountability and trust.

- Performance and reliability risk
  The risk that models underperform, drift over time, or behave unpredictably in edge cases, leading to wrong or inconsistent decisions.

- Security and misuse risk
  The risk that AI systems are attacked (e.g., data poisoning, prompt injection, model theft) or misused (e.g., used beyond the intended scope) to harmful effect.

- Privacy and data protection risk
  The risk that AI systems use data in ways that violate privacy expectations, consent, or legal constraints, or expose sensitive data through outputs.

- Reputational and societal risk
  The risk that AI usage damages public trust, harms vulnerable populations, or makes the organization appear irresponsible or out of step with social expectations.

These categories provide a checklist: every AI initiative should be examined against all of them. Some will be more relevant than others depending on the use case, but none should be ignored.

Diagram 6.1: The AI Risk Wheel

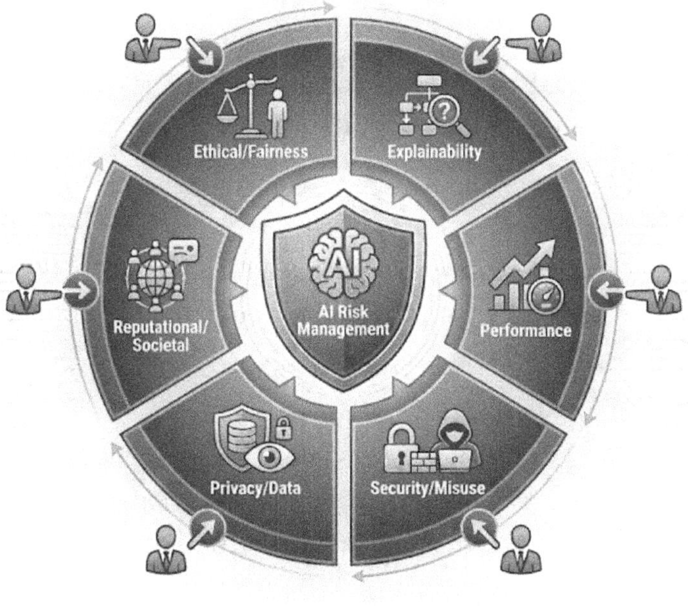

6.4 Integrating AI risk into enterprise risk management

AI risk should not sit in a silo. It must be integrated into your **enterprise risk management** (ERM) framework so that AI-related issues are visible at the same level as other strategic and operational risks. That means AI risk needs to appear in risk registers, be discussed in risk committees, and be considered in audits and assurance activities.

Practically, this involves:

- Adding AI-specific risk categories to your risk taxonomy
  Ensure that your risk language and categories explicitly include AI and algorithmic risks, not just generic technology risks.

- Mapping AI use cases to existing risk domains
  For example, an AI system for credit decisions links to credit risk, conduct risk, and reputational risk; an AI system for safety monitoring links to operational and safety risk.

- Defining risk appetite for AI
  Clarify how much AI-related risk is acceptable in different areas (e.g., low tolerance in safety-critical domains, moderate in internal efficiency tools).

- Including AI scenarios in risk assessments and stress tests
  Consider questions like "What if this AI system fails in a particular way?" and "What if media or regulators challenge this use of AI?"

- Reporting AI risk indicators to leadership
  Develop simple metrics, such as the number of high-risk AI systems, incidents, exceptions, or overdue reviews, and include them in enterprise risk dashboards.

When AI risk is integrated into ERM, risk officers and executives see it, understand it, and can prioritize responses. Leaving AI risk outside of ERM guarantees it will be under-resourced and reactive.

6.5 Compliance in the AI era: beyond ticking the box

Compliance in the context of AI is not just about avoiding fines or legal sanctions. It is about demonstrating that you understand and manage the risks of AI in a way that respects laws, regulations, and internal policies. As AI regulations and guidance evolve, they increasingly focus on processes, documentation, and accountability rather than prescribing specific algorithms or tools.

For managers, this translates into three key compliance expectations:

- Demonstrable governance
  You must be able to show that AI systems have gone through defined governance steps: risk assessment, testing, approval, monitoring, and periodic review.

- Traceability and documentation
  You should have up-to-date documentation for each

significant AI system, including purpose, data sources, risk tier, tests performed, results, owners, and change history.

- Responsiveness and remediation
  You need processes to respond to incidents or findings quickly—adjusting models, updating policies, and communicating with stakeholders where necessary.

Meeting these expectations is not about producing a single compliance report. It is about having **living evidence** that your AI practices are responsible and controlled. Your AI governance framework from earlier chapters is the backbone that makes this possible.

Diagram 6.2: AI Risk and Compliance Funnel

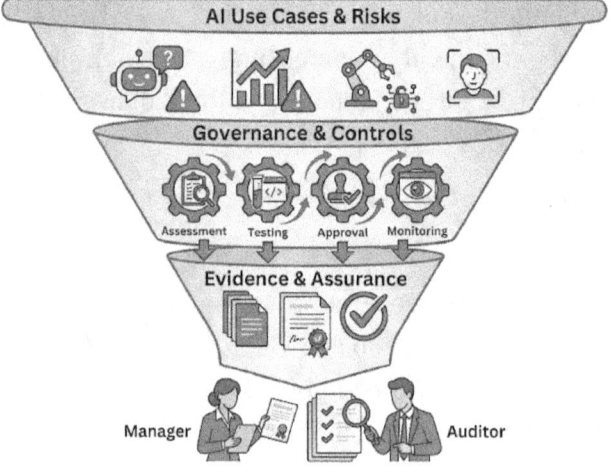

6.6 Responsible AI: turning principles into operational practices

"Responsible AI" is a phrase used widely. Still, it only has real meaning when principles are translated into operational practices—things people actually do, decisions they document, and controls they apply. You can think of Responsible AI as the **value-and-ethics layer on top of your risk-and-compliance** efforts.

Operational Responsible AI practices include:

- Explicit fairness goals and tests
  Rather than hoping for fairness, define which fairness metrics matter (e.g., similar error rates or approval rates for comparable groups) and test for them.

- Human-in-the-loop and human-on-the-loop designs
  Define when and how human oversight is required, ensuring humans can understand, override, or stop AI behavior when needed.

- Transparency for affected stakeholders
  Provide clear communication about where AI is used, what it does, and how people can seek clarification or contest decisions.

- Stakeholder engagement
  For high-impact AI systems, involve representatives of affected groups (e.g., patients, citizens, employees) in design or review discussions when feasible.

- Ongoing ethical review
  Periodically revisit whether an AI system remains appropriate, especially as contexts or societal expectations change.

Embedding these practices turns Responsible AI from a slogan into a reality of governance. They also reduce the risk of surprises: if you have continuously considered ethics and stakeholder impact, you are less likely to encounter severe backlash later.

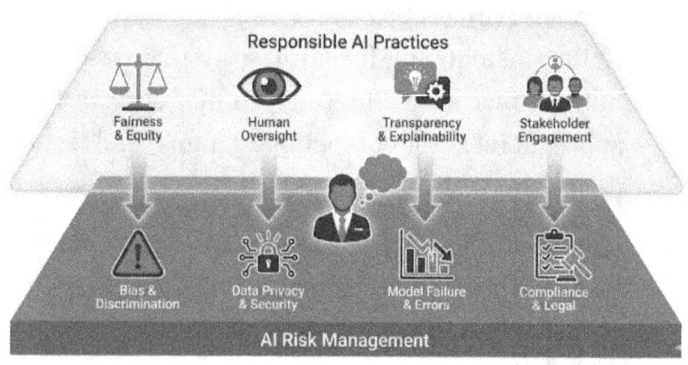

Diagram 6.3: Responsible AI as an Overlay on Risk

6.7 The AI risk assessment template: what to include

To make AI risk, compliance, and responsibility concrete, you need a standard **AI risk assessment template**. This becomes a core artifact in your governance framework—used during intake, design, and approval. Key sections typically include:

- Use case overview
  Description of what the AI system will do, who it affects, and how it will be used in workflows.

- Risk tier classification
  Identification of the risk tier (e.g., minimal, moderate, high, critical) and rationale based on impact and context.

- Data and model summary
  Data sources, key features, model type, and any third-party components or services used.

- Risk category assessment
  Structured questions and ratings for fairness, explainability, performance, security, privacy, and reputational risks, including mitigations planned.

- Controls and oversight plan
  Defined controls (e.g., testing, human oversight, monitoring, review frequency) are tied to the risk tier.

- Approval and accountability
  Named owners, sign-offs (e.g., business sponsor, IT, risk, compliance), and conditions or limitations on use.

As a manager, you should ensure that every significant AI system has a completed and current version of this template. Over time, it becomes a powerful repository for audits, reviews, and organizational learning.

6.8 The manager's checklist: stress-testing AI risk and compliance readiness

Use the following questions to test how ready your organization is to manage AI risk and demonstrate compliance:

- Do we have an agreed list of AI risk categories and a shared language for them?
  If different teams use different terms, risk conversations will be fragmented and confusing.

- Are AI initiatives consistently assigned risk tiers, and do those tiers drive different levels of review?
  If every AI project is treated the same, you either over-govern or under-govern many of them.

- Can we point to a standard AI risk assessment template and show completed examples?
  If templates exist but are rarely used or incomplete, they are not yet part of real practice.

- Do our existing risk and compliance functions know how AI fits into their mandate?
  If AI feels "new" to them and is not covered in their processes, you have a coordination gap.

- When an AI-related concern arises (e.g., a fairness complaint), do we have a documented path to investigate and respond?
  If the response depends on ad hoc decision-making, you are exposed.

Each "no" or "not sure" is not a failure; it is a map of where to improve. Your role is to turn these questions into concrete work items and timelines.

6.9 AI Reality Check: You can't eliminate risk, but you can choose it wisely

A common misconception is that the goal of AI governance is to eliminate all risk. In reality, that is neither possible nor desirable. Every meaningful innovation carries some risk; the real question is whether the risk is **understood, consciously accepted, and properly controlled**. Trying to eliminate all risk can lead to paralysis, shadow AI, or superficial compliance that hides real issues.

A more mature stance sees AI risk management as a form of **informed choice**. You identify where AI can create real value, understand the associated risks, design controls that are proportionate, and then make a conscious decision to proceed or not. You accept that some residual risk remains, but you actively monitor and manage it. When something goes wrong, you are prepared: you have evidence of due diligence, mechanisms for remediation, and a culture that learns rather than panics.

As a manager, your goal is not to say "no" to AI. It is to say "yes, under these conditions." Responsible AI governance is what defines those conditions in a clear, repeatable way.

6.10 How this chapter connects to your governance journey

In this chapter, you connected AI governance to risk, compliance, and responsible practice. You defined an AI risk landscape, saw how to integrate AI risk into enterprise risk management, and explored how compliance increasingly focuses on evidence of governance rather than just outcomes. You also learned how Responsible AI practices overlay and strengthen your risk and compliance work, and how templates and checklists translate these ideas into operational reality.

In the next chapters, you will deepen this journey. Chapter 7 will focus on **Operationalizing Ethics and Transparency**, translating high-level values into day-to-day practices, communications, and tools. Chapter 8 will take those concepts into complex environments—**Cloud and Hybrid AI**—where shared responsibility with vendors and partners becomes a central governance concern. Step by step, you are building a comprehensive, manager-ready approach to AI governance that spans strategy, operations, risk, ethics, and technology.

7 Operationalizing Ethics and Transparency

7.1 Opening vignette: "The policy everyone signed and no one followed."

A major organization proudly published its Responsible AI Policy on the intranet. The document looked impressive: it referenced fairness, transparency, accountability, and human-centric design. Leaders announced it in an all-hands meeting and asked managers to "make sure your teams follow this." For a few weeks, people mentioned the policy in presentations and emails. Then work went back to normal. AI projects continued, but the policy rarely appeared in design discussions, sprint planning, or vendor evaluations.

Months later, a controversy emerged around an AI-powered recommendation system that appeared to disadvantage a specific group. When leadership asked, "Did this system comply with our Responsible AI Policy?" no one could answer confidently. The product team said they had "considered ethics," but had no documentation to support it. Data scientists mentioned they "checked for bias," but there were no standardized tests or thresholds in place. Legal had not reviewed the system because it was treated as a minor feature. The gap was clear: the **policy existed on paper, but ethics and transparency were not operationalized**.

This chapter is about closing that gap. It explains how to turn lofty ethical principles into concrete actions, tools, and

99

habits that shape how AI is designed, built, deployed, and communicated—every day.

7.2 Why ethics must move from principle to practice

Ethical aspirations are necessary, but not sufficient. Without practical mechanisms, ethics becomes a poster on the wall rather than a property of your systems. For AI, this is particularly dangerous. AI systems:

- Scale fast, so small ethical oversights can become large, systemic issues.
  A biased rule in a spreadsheet can affect dozens of cases; a biased AI model can affect thousands or millions.

- They are often opaque, making it hard to see ethical problems early.
  If no one is explicitly looking for fairness or explainability gaps, they go unnoticed until a complaint or incident.

- Influence human behavior and institutional norms.
  Once AI is embedded in workflows, people may defer to it, assuming the system has been "ethically vetted" even when it hasn't.

To avoid this, ethics must be woven into the routines of work: the questions teams ask, the artifacts they create, the reviews they conduct, and the way they communicate with stakeholders. Ethics becomes a **way of working** rather than an afterthought. This chapter focuses on giving you tools to make that shift as a manager.

7.3 The ethics-to-workflow chain

A useful mental model is the **ethics-to-workflow chain**: how a high-level principle becomes a practical behavior. You can think about it in four stages:

1. Principle
   A short statement of value (e.g., "We will avoid unfair bias in AI systems.").

2. Operational standard
   A specific expectation (e.g., "All high-risk AI systems must be tested for bias using agreed metrics and thresholds.").

3. Workflow step
   A concrete action in a process (e.g., "Run fairness tests and document results before design review.").

4. Artifact or tool
   A tangible output (e.g., a bias report, dashboard screenshot, or checklist entry).

As a manager, your task is to ensure the chain is complete. If your principles do not align with standards, if standards do not align with workflows, or if workflows do not result in artifacts, ethics will not be truly operational.

From Principle to Practice

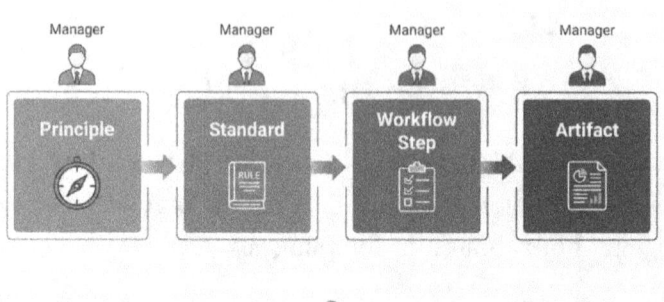

7.4 Translating common AI ethics principles into concrete actions

Most organizations converge around a similar set of AI ethics principles: fairness, transparency, accountability, privacy, and human-centric design. The challenge is defining what these mean **in practice**. Below are manager-ready translations you can adopt or adapt.

- Fairness
  Practical actions: identify impacted groups, define fairness metrics relevant to the use case, run subgroup performance tests, and document trade-offs when perfect fairness is unattainable.

- Transparency
  Practical actions: label where AI is used in user-facing interfaces, create internal fact sheets or "model cards," and ensure teams can explain inputs, outputs, and limitations in plain language.

- Accountability
  Practical actions: assign named owners for each AI system, document decisions at key milestones, and define incident response paths for concerns or complaints.

- Privacy and respect for persons
  Practical actions: confirm data purposes align with user expectations, avoid using sensitive attributes unless justified and governed, and implement privacy-aware logging and monitoring.

- Human-centric design
  Practical actions: involve frontline users in design and testing, make interfaces that support human judgment (not just automation), and design clear override and feedback mechanisms.

Each principle should have a **short list of required actions** for high- and medium-risk AI systems. That list becomes part of your project templates and review checklists.

7.5 The ethics review is part of existing governance, not an add-on

Ethics reviews often fail when they are bolted onto the end of a project as a "sign-off." By then, timelines are tight, decisions are locked in, and teams see ethics as a hurdle rather than a design input. To succeed, ethics and transparency checks must be embedded into existing governance steps you already use.

For example:

- During use case intake:
  Include questions about potential impacts on vulnerable groups, rights, and trust, not just value and feasibility.

- During design reviews:
  Require a brief ethics and transparency section: what are the main ethical considerations and how are they addressed? How will users know AI is involved?

- During testing:
  Make fairness and explainability tests part of the standard test suite for higher-risk AI systems.

- During deployment approval:
  Confirm that required ethical artifacts (e.g., bias tests, user communication plan, oversight design) are completed and attached.

By integrating ethics and transparency into these existing steps, you avoid creating a separate "ethics bureaucracy." Instead, you enhance workflows your teams already understand.

Diagram 7.2: **Embedding Ethics into the AI Lifecycle**

7.6 Tools for fairness and bias assessment (at a manager level)

You do not need to pick fairness algorithms yourself, but you should ensure that **bias assessment is a standard practice** where appropriate. Key manager-level expectations:

- For high- and some medium-risk AI systems, require:

    o Identification of relevant groups or segments (e.g., age ranges, regions, income brackets).

    o Performance breakdowns across these groups (e.g., error rates, false positives, false negatives).

- o Documentation of any significant disparities and mitigation steps taken.
- Ask your teams:
  - o "Which groups might be harmed if this model behaves unfairly?"
  - o "What fairness metrics are we using, and why?"
  - o "What did we learn from subgroup analysis, and what did we change as a result?"
- Embed fairness into monitoring:
  - o Track key metrics by relevant segments over time, not only in aggregate.

To make this practical, your organization can develop standardized **fairness test templates** and dashboards for common use case types (e.g., risk scoring, prioritization, recommendations). Your role is to insist these exist and are used, especially in higher-risk domains.

7.7 Transparency in practice: what to communicate with and to whom

Transparency operates at multiple levels:

- Internal transparency
  What your teams and leaders know about AI systems. This includes documentation of purpose, design, data, performance, and limitations.

- Operational transparency
  What frontline staff and operational managers know about how AI impacts their workflows—where AI is embedded, how they should interpret its outputs, and when to override.

- External transparency
  What end users, customers, or citizens know about AI's role in decisions that affect them, and how they can seek clarification or challenge outcomes.

As a manager, you should ensure that for each significant AI system:

- There is a clear internal "fact sheet" or model summary that non-technical leaders can read.

- Frontline staff receive targeted training and reference guides explaining how AI fits into their work.

- External communications (where appropriate) clearly indicate where AI is used and how individuals can get more information or raise concerns.

Transparency is not about exposing source code or proprietary models; it is about giving people enough information to understand and meaningfully engage with AI decisions.

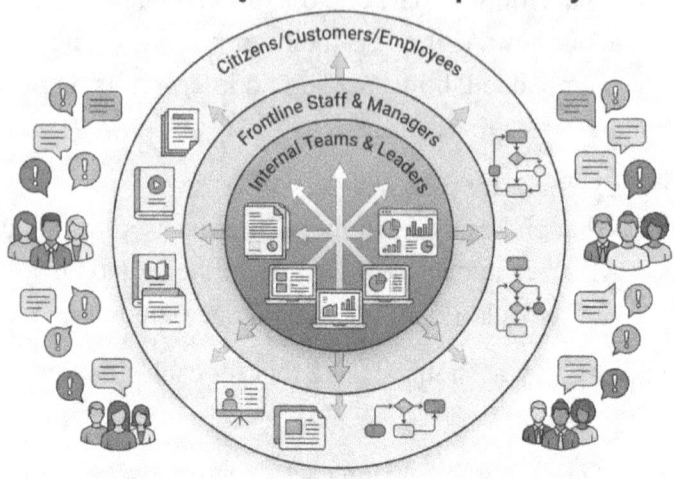

7.8 Building ethics and transparency into vendor relationships

Many AI capabilities enter your organization through vendors—SaaS applications, cloud services, and packaged AI components. Operationalizing ethics and transparency means extending your expectations to these third parties.

Practically, this means:

- Including ethics and transparency requirements in procurement and RFPs
  Ask vendors to describe how they test for bias, how they document models, and what explainability features they provide.

- Requiring minimum documentation
  At least a high-level description of the model, data sources, key limitations, and known risk areas, written in understandable language.

- Negotiating transparency and audit rights
  Where risk is high, ensure you can access sufficient information (or third-party certifications) to evaluate ethical risks and compliance, even if the underlying model is proprietary.

- Defining shared responsibilities
  Clarify which aspects of ethics and transparency the vendor handles (e.g., core algorithm design) and which you handle (e.g., how the system is used, how users are informed).

As a manager, your role is to ensure vendor governance is not just about price and SLA, but also about **responsible AI criteria** aligned with your internal standards.

7.9 The manager's ethics-and-transparency checklist

To make this chapter actionable, use the following checklist when reviewing or sponsoring AI initiatives:

- Have we identified the main ethical questions and impacted groups for this use case?
  If not, the project is at risk of overlooking key fairness and harm considerations.

- Are fairness and transparency requirements documented alongside functional requirements? If ethics is not in the requirements, it will be an afterthought in design.

- Do we have at least one concrete fairness test or analysis scoped for higher-risk AI? If tests are vague ("we'll check later"), they rarely happen.

- Is there a clear explanation of the system that a non-technical leader can understand? If such an explanation does not exist, internal transparency is weak.

- Do frontline staff know when AI is involved, how to interpret its outputs, and how to escalate concerns? If they are unsure, your operational transparency and ethics-by-design are not yet complete.

You can embed these questions into your governance templates, steering committee materials, or project gate reviews. Over time, they will shape how teams plan and execute AI work.

7.10 AI Reality Check: Ethics is not about perfection

A common stumbling block is the belief that ethics requires perfect fairness, perfect transparency, or perfect alignment with every stakeholder's expectations. That bar is impossible to meet and can lead to paralysis or purely symbolic efforts. In reality, ethical AI is about **better decision-making and fewer avoidable harms**, not perfection.

You will face trade-offs. Improving one fairness metric may worsen another. Providing more transparency may conflict with security or privacy constraints. Some stakeholders will disagree about what is "fair" or "acceptable." The goal of operationalizing ethics and transparency is to make these trade-offs **explicit, documented, and revisitable**, rather than implicit and invisible.

As a manager, your job is to create a culture where teams can surface ethical tensions early, discuss them openly, and document decisions, rather than pushing them aside. When the inevitable questions come—from leadership, oversight bodies, or the public—you will then have a record showing that you took ethics seriously and acted in good faith within real-world constraints.

7.11 How this chapter advances your governance journey

In this chapter, you moved from AI ethics as a high-level aspiration to an **operational practice**. You learned how to connect principles to standards, workflows, and artifacts, and how to embed ethics and transparency into existing governance steps. You explored concrete ways to assess fairness, communicate about AI systems, and extend ethical expectations to vendors. You also gained a practical checklist to keep ethics and transparency present in your daily decisions.

Next, in Chapter 8, we will focus on **Governance in the Cloud and Hybrid AI Environments**, where responsibilities are shared among internal teams and external providers. You will see how to apply the ethical and transparent practices you have learned in more complex technical and organizational landscapes, ensuring that governance remains strong even as AI spans multiple platforms and partners.

8 Governance in the Cloud and Hybrid AI Environments

8.1 Opening vignette: "It's in the cloud, so it's their problem… right?"

An IT manager overseeing a major modernization program proudly reported that most of the organization's AI capabilities now ran in the cloud. Core applications had been migrated to a leading provider, and several new AI services were integrated through APIs—text analytics, anomaly detection, and a conversational assistant. The manager took comfort in the provider's marketing: "enterprise-grade security," "responsible AI," and "compliance-ready." When internal auditors asked about AI governance, the answer was simple: "The cloud vendor handles that."

Trouble arrived when a regional regulator requested evidence of how a specific AI service—used to classify incoming citizen requests—was being governed. The regulator wanted to know what data was being sent to the cloud, where it was stored, who could access it, and how the model's behavior was monitored. The vendor provided generic documentation, but not enough detail to answer questions about this organization's specific configuration, data flows, or oversight. The regulator's message was clear: "You cannot outsource accountability." The manager realized that while infrastructure had moved to the cloud, **governance had not followed**.

This chapter explains how to fix that. It shows you how to govern AI in cloud and hybrid environments, where responsibilities are shared between your organization and external providers.

8.2 Why cloud and hybrid AI complicate governance

Cloud and hybrid environments introduce both power and complexity. On the positive side, cloud platforms provide scalable compute, advanced AI services, and managed tooling that would be difficult or expensive to build in-house. On the challenging side, they distribute responsibilities across organizational boundaries. Data may reside in multiple regions, providers may update models, and AI services may be "black boxes" you consume rather than code you control.

This complexity creates governance challenges:

- Visibility
  You may not see all internal and external AI services in use, especially if teams can easily provision cloud resources.

- Control
  You control configurations, data inputs, and usage patterns, while the provider controls infrastructure, base models, and many security measures.

- Accountability
  Stakeholders hold **you** accountable for AI outcomes,

even when parts of the stack are outside your direct control.

As an IT manager, you must design governance that acknowledges this shared responsibility, clarifies who does what, and ensures that your obligations—to your mission, stakeholders, and regulators—are met regardless of where the technology runs.

8.3 Shared responsibility for AI: who owns what

Most cloud providers use a "shared responsibility model" for security and compliance. For AI, you need to adapt that concept to include governance, ethics, and risk. Put simply, providers are responsible for the **platform**; you are responsible for the **use**.

You typically remain responsible for:

- Deciding which AI services to use, and for what purposes.

- Choosing and managing the data you send to cloud AI services.

- Configuring models, prompts, access controls, and logging.

- Ensuring AI usage aligns with laws, policies, and ethical commitments.

- Monitoring AI behavior in your workflows and responding to issues.

Providers are typically responsible for:

- Securing the underlying infrastructure and core platform.

- Managing base models and core service updates.

- Providing configuration options, logging features, and security controls.

- Publishing documentation and compliance attestations for their services.

Your governance must make this division explicit, not assumed. That means capturing it in policies, contracts, architectures, and operating procedures, so teams understand where provider responsibilities end, and yours begin.

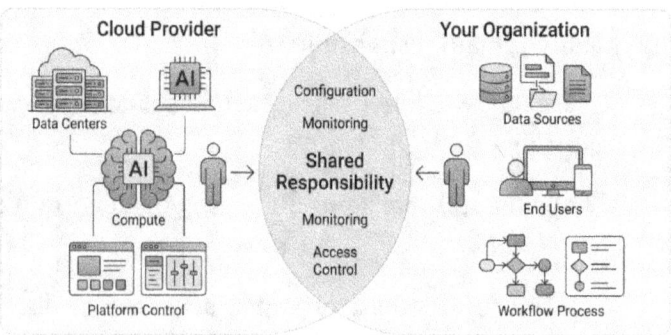

Shared Responsibility for Cloud AI

8.4 Governance objectives in cloud and hybrid AI

In cloud and hybrid settings, your AI governance has five main objectives:

- Visibility
  Knowing which AI services, models, and providers are in use, where, and for what.

- Consistency
  Applying your governance principles, risk tiering, and controls consistently across on-prem, cloud, and hybrid environments.

- Control over data
  Ensuring that data movement, storage locations, and usage are compliant and aligned with your risk appetite and obligations.

- Control over configuration and integration
  Ensuring that how you configure and embed cloud AI into workflows reflects your governance standards.

- Assurance
  Being able to demonstrate to leadership, auditors, and regulators that you understand and manage AI risks in these environments.

These objectives help you prioritize governance efforts. You do not need to control everything in the cloud, but you do need to **control enough** to meet these objectives reliably.

8.5 Cloud AI services: key questions to ask before adoption

Before you adopt a cloud AI service—such as text classification, translation, vision APIs, or general-purpose models—your governance process should include targeted questions that reflect cloud specifics. At a minimum, ask:

- What data will we send to this service, and is it appropriate to send it to the cloud? Consider sensitivity, privacy, regulatory constraints, and whether anonymization or minimization is needed.

- Where will the data be stored or processed, geographically and logically? Data residency and cross-border transfer rules may apply, particularly in regulated sectors.

- What logs will be generated, and who can access them?
  Logs may contain sensitive information; you need to control and monitor access to them.

- How is the model managed by the provider (updated, retrained, versioned)?
  Frequent updates may be beneficial, but they can also change behavior; you need to know how and when this happens.

- What configuration options exist (e.g., safety filters, content limits, prompt controls)?
  These options often implement provider-level

governance; you must set them in line with your risk appetite.

- What compliance and security assurances does the provider offer, and do they meet your needs? Look for certifications, documentation, and contractual commitments that align with your obligations.

By standardizing these questions (and their answers) in your intake and architecture review templates, you make cloud AI adoption more predictable and controlled.

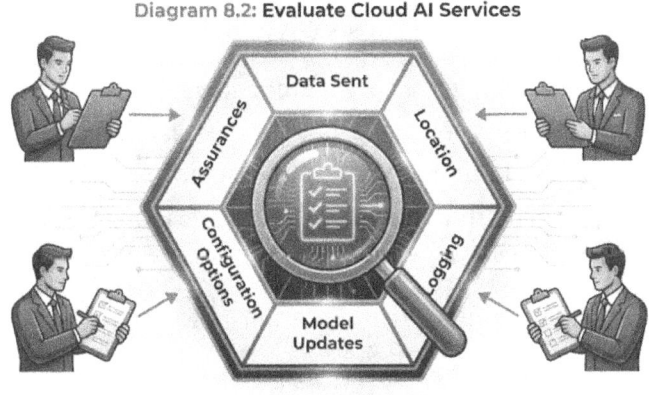

Diagram 8.2: **Evaluate Cloud AI Services**

8.6 Hybrid AI architectures and their governance implications

Many organizations end up with hybrid AI architectures: some models and data on-premises, others in one or more clouds, and still others embedded within SaaS applications. This can be technically efficient but governance challenging.

Typical hybrid patterns include:

- **On-prem data, cloud models**
  Sensitive data stays on-premises, while de-identified or transformed data is sent to cloud models. Governance focus: data transformation, anonymization, and accurate representation.

- **Cloud data lake, mixed model locations**
  Data resides in a cloud lake; some models run in the same cloud, others on-prem or in different clouds. Governance focus: data access, cross-cloud movement, and consistent policies.

- **SaaS applications with embedded AI**
  Vendors incorporate AI features that use your data but are largely opaque. Governance focus: understanding how the vendor uses your data and configuring features responsibly.

In each pattern, your governance must clarify:

- Where the "system of record" for data and models resides.

- Which environments are authoritative for monitoring and logging?

- What policies apply across environments (e.g., retention, access, masking).

Hybrid does not excuse inconsistency; it increases the need for clear, cross-environment governance rules.

Hybrid AI Topology and Governance Points

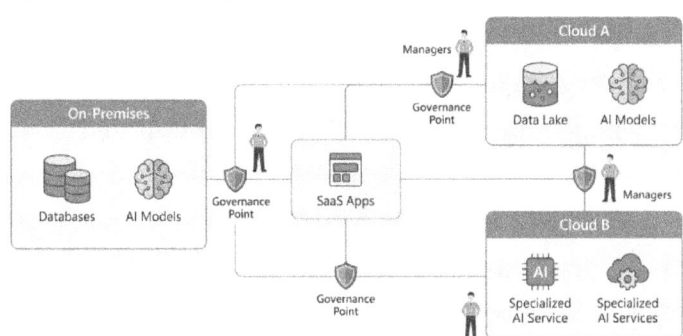

8.7 Controlling data in cloud and hybrid AI

Data control is central to governance in cloud and hybrid AI environments. You must ensure that data handling aligns with your policies and external obligations, regardless of where the data physically resides. Key practices include:

- Data classification
  Classify data by sensitivity and apply rules for which classes can be used with which AI services (e.g., "no highly sensitive data into generative AI services without explicit approval").

- Data minimization and anonymization
  Send only the data necessary for the AI purpose; wherever possible, remove direct identifiers or reduce detail to lower risk.

- Location and residency controls
  Use provider options to keep data in approved regions and ensure your contracts reflect residency requirements.

- Access control and segregation
  Implement fine-grained access controls and separation of environments (e.g., dev/test vs production) across cloud and on-prem.

- Logging and retention policies
  Define which logs to keep, where to keep them, for how long, and how they are protected, given that logs may contain sensitive data.

These practices should be codified in your data and AI governance policies and enforced through technical controls, including identity and access management, network segmentation, data loss prevention, and logging configurations.

8.8 Integrating cloud providers into your governance framework

Governance in cloud and hybrid environments is not only a technology issue; it is a relationship issue. Providers and vendors are part of your extended operating model. To integrate them into your governance framework:

- Extend your risk tiering to include vendor and service risk
  Higher-risk AI use cases that rely on external

services should trigger more thorough vendor due diligence and tighter contractual controls.

- Define provider governance criteria
  For example, the minimum documentation, security practices, data handling commitments, and ethical AI posture you expect.

- Use standardized vendor evaluation and review processes
  Treat AI providers like other critical vendors: with periodic risk reviews, performance assessments, and oversight.

- Align incident response
  Ensure your incident management plan includes scenarios where the issue lies partly or entirely with a provider, including contact paths, responsibilities, and communication expectations.

- Maintain an inventory of AI-relevant providers
  Know who they are, what services they provide, and where they intersect with your data and critical workflows.

This integration ensures that "cloud" does not become a governance black box, but a well-understood and actively managed part of your AI ecosystem.

Diagram 8.4: Provider in the AI Governance Framework

8.9 Monitoring AI in cloud and hybrid environments

Monitoring AI in cloud and hybrid environments is more than checking uptime. You must monitor **behavior**: performance, fairness, drift, security events, and usage patterns. Challenges arise because telemetry is scattered across your systems, provider dashboards, and sometimes embedded in SaaS apps.

Practical steps:

- Define a minimum monitoring baseline for all significant AI systems
  Metrics might include accuracy proxies, error rates, usage volume, override rates, and incident counts.

- Integrate logs and metrics
  Where possible, centralize key logs and metrics into your monitoring and observability tools, even if they originate in provider platforms.

124

- Monitor for misuse and unexpected patterns
  For generative services, track signs of prompt misuse, data leakage, or unexpected content types.

- Set thresholds and alerts
  Define what constitutes a notable deviation (e.g., a surge in errors or a sudden change in outputs) and configure alerts to reach the right teams.

- Periodically review monitoring data in governance forums
  Bring AI monitoring results to risk and governance committees so leadership can see trends and act on them.

Monitoring is your early warning system; without it, AI in the cloud can diverge from expectations without anyone noticing until external stakeholders do.

8.10 The manager's checklist: cloud and hybrid AI governance readiness

Use this checklist to assess your readiness to govern AI in cloud and hybrid environments:

- Do we have a current inventory of AI services and models, including cloud and SaaS-based ones?
  If you cannot list what exists, you cannot govern it.

- For each major AI service, do we know what data it receives, where data is processed, and what logs are generated?

If this information is unknown or scattered, your data and accountability posture are weak.

- Is the shared responsibility for each cloud AI service documented and understood (what we do vs what the provider does)?
  If responsibilities are assumed but not written, gaps will emerge.

- Are AI-specific requirements (risk, ethics, transparency) embedded in vendor selection and contract management?
  If contracts focus only on cost and uptime, AI-specific risks are unmanaged.

- Do we have integrated monitoring and incident response that covers issues arising in cloud and hybrid AI?
  If cloud incidents rely solely on provider notifications, you have limited control.

These questions should feed into your broader cloud governance and AI governance improvement plans.

8.11 AI Reality Check: "cloud-first" doesn't mean "governance-optional."

"Cloud-first" strategies can inadvertently foster the belief that providers are responsible for everything important. For AI, this belief is not just wrong; it is dangerous. Providers can give you tools, safeguards, and documentation. They cannot decide for you what is ethical, what is acceptable risk, or how AI should be used in your unique mission context.

A mature stance treats the cloud as an **amplifier** of your governance—allowing you to implement policies and controls faster and at scale—but not as a substitute for governance. You still define the rules, choose what to build, configure services responsibly, and remain answerable to your leadership, stakeholders, and regulators. "Cloud-first" must be matched with "governance-first" thinking.

8.12 How this chapter advances your governance journey

In this chapter, you extended your AI governance approach to cloud and hybrid environments. You learned how shared responsibility works for AI, how to evaluate and control cloud AI services, and how to manage hybrid architectures without losing consistency. You saw how to integrate providers into your governance framework, centralize monitoring, and maintain accountability even when much of the stack sits outside your data center.

The remaining chapters build on this foundation. Chapter 9 will focus on **Governance for Generative AI and Autonomous Systems**, where content generation, autonomy, and human-in-the-loop dynamics create new governance challenges. Chapter 10 will dive deeper into **People, Skills, and Governance Culture**, ensuring that your governance is not merely structural but truly embedded in how people think and act. Step by step, you are constructing a comprehensive, manager-ready AI governance playbook that works across technologies, environments, and organizational boundaries.

9 Governance for Generative AI and Autonomous Systems

9.1 Opening vignette: "It wrote the email, but we owned the fallout."

A customer service director greenlit a pilot of a generative AI assistant to help agents draft responses to complex complaints. The system was configured to propose full email drafts based on case notes, previous correspondence, and policy documents. Early feedback was enthusiastic: agents loved the speed boost and appreciated that the assistant "knew" the knowledge base better than most new hires.

Then a complaint escalated. A customer received a technically accurate but cold and dismissive response that cited an internal policy in a way that made the organization seem uncaring and inflexible. The customer shared the email publicly, and it went viral. Leadership demanded to know who wrote it. The agent said, "The AI drafted it; I just clicked send." The AI vendor noted that the organization controlled the prompts and approvals. Legal noted there was no clear guidance on when agents could modify or override AI-generated content. The problem was not just the model; it was the lack of **governance for generative AI** and the blurred line between assistance and delegation.

This chapter addresses that gap. It explains how to govern generative AI and autonomous systems so that people understand their responsibilities, risks are consciously managed, and your organization can benefit from these powerful tools without losing control.

9.2 Why generative and autonomous AI raise the stakes

Generative AI and autonomous systems differ from many earlier AI applications in ways that matter for governance:

- They create content, not just scores or labels.
  Outputs are often human-readable text, images, code, or actions that directly shape stakeholder experiences.

- They can "hallucinate" or fabricate plausible but false information.
  This makes blind trust dangerous, especially in high-stakes communications or decisions.

- They can operate semi- or fully autonomously.
  In autonomous workflows (e.g., automated approvals, robotic process automation, or agentic systems), human oversight can be thin or delayed.

- Prompts, context, and configuration heavily influence their behavior.
  Governance must consider not only the base model, but how it is used, combined with tools, and integrated into workflows.

Because of these properties, generative and autonomous systems demand **sharper lines of accountability**, **clearer**

usage boundaries, and **stronger oversight patterns** than many predictive models.

9.3 Key governance questions for generative and autonomous systems

When generative or autonomous AI is proposed, you should consistently ask:

- What decisions or actions will the system take, and which remain with humans?

- What kinds of content will it produce, and who will see or depend on that content?

- What is the potential impact if the system makes a plausible but incorrect or harmful output?

- How will people know that AI is involved, and how can they correct or override it?

- What safeguards (technical and process) are in place to limit unsafe or inappropriate outputs?

These questions move governance beyond "Can we connect to this model?" toward "Should we, and under what conditions?"

Governance Lens for Generative and Autonomous AI

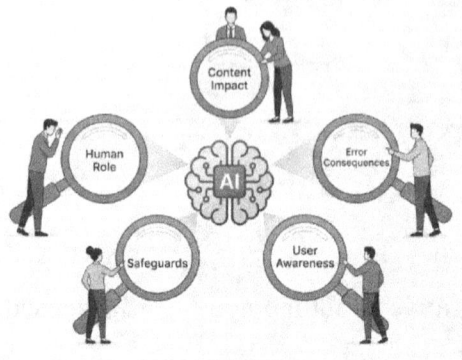

9.4 Levels of autonomy: assist, augment, act

Governance improves when you clearly define the **level of autonomy** for each AI system. A practical classification:

- Assist
  AI suggests content or options, but humans decide and act (e.g., draft emails, suggested responses, recommended actions).

- Augment
  AI performs tasks with human oversight and periodic review (e.g., pre-sorting cases, drafting policy summaries, ranking recommendations). Humans usually accept outputs but can intervene when necessary.

- Act
  AI initiates actions autonomously within defined boundaries (e.g., auto-approving low-risk claims,

auto-routing work, triggering alerts with follow-on actions). Human involvement is limited or after the fact.

Each level requires different governance:

- Assist-level systems: focus on user training, clear UI signals, and guidance on when to edit or reject AI content.

- Augment-level systems: add monitoring of acceptance rates and error patterns, plus clear policies on when humans must review.

- Act-level systems: require rigorous risk assessment, strong guardrails, audit trails, and clear escalation paths when thresholds or anomalies are detected.

By explicitly labeling each system with its autonomy level, you help teams select appropriate controls and avoid accidental drift from assistive to autonomous use.

Diagram 9.2: Autonomy Ladder and Governance Intensity

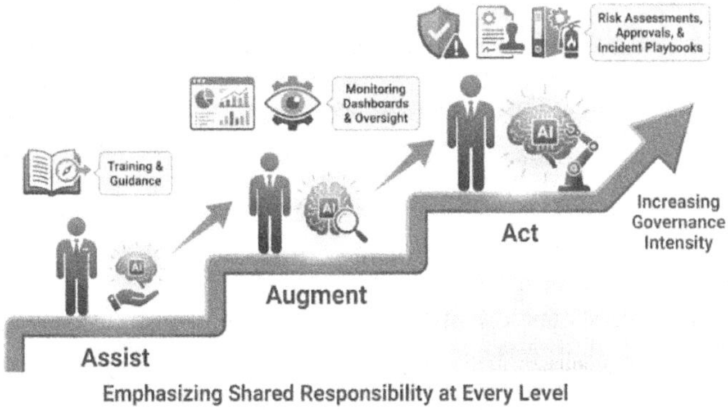

Emphasizing Shared Responsibility at Every Level

9.5 Guardrails for generative AI: prompts, filters, and boundaries

Generative models are highly configurable through prompts, system instructions, and integration patterns. Governance must define **guardrails** that constrain behavior.

Key guardrail dimensions:

- Input restrictions
  Define what data can be fed into generative systems (e.g., no sensitive personal data without explicit approval, no proprietary details in public models).

- Prompt and system message design
  Standardize prompt patterns that emphasize accuracy, caution, adherence to policy, and refusal to answer outside scope. Avoid ad-hoc prompts that invite risky behavior.

- Output filters and policies
  Configure safety settings where available (e.g., content filters, blocked categories) and define content rules (e.g., no legal advice, no medical diagnosis without human validation).

- Role-based capabilities
  Differentiate what the AI can do for different user roles (e.g., internal staff vs public-facing chatbots) and attach appropriate limits and logging.

- Integration boundaries
  Restrict where outputs can flow automatically; for example, require human approval before AI-

generated content is sent externally or used to trigger significant actions.

These guardrails should be encoded in both technical configuration and written policies so they can be understood, implemented, and audited.

9.6 Human-in-the-loop and human-on-the-loop patterns

Generative and autonomous systems must be designed with clear **human oversight patterns**:

- Human-in-the-loop
  Humans review, approve, or edit AI outputs before they affect external stakeholders (e.g., agents editing AI-drafted messages before sending). Governance focus: training, workload, and expectations for edits.

- Human-on-the-loop
  AI operates, but humans oversee performance and can intervene based on dashboards or alerts (e.g., autonomous routing or approvals with monitoring). Governance focus: thresholds, review frequency, and escalation paths.

- Human-out-of-the-loop (only in narrow, controlled contexts)
  Fully automated actions with only periodic audits. Suitable only for low-risk, well-understood scenarios with strong guardrails. Governance focus: rigorous pre-deployment assessment and strong monitoring.

As a manager, you should insist that each generative or autonomous system explicitly specify:

- Which oversight pattern applies?

- What exact actions humans must take.

- How those humans are informed and trained.

- How the system escalates when behavior deviates from norms.

Without this, "human oversight" easily deteriorates into a checkbox rather than a real safeguard.

Oversight Modes for AI Systems

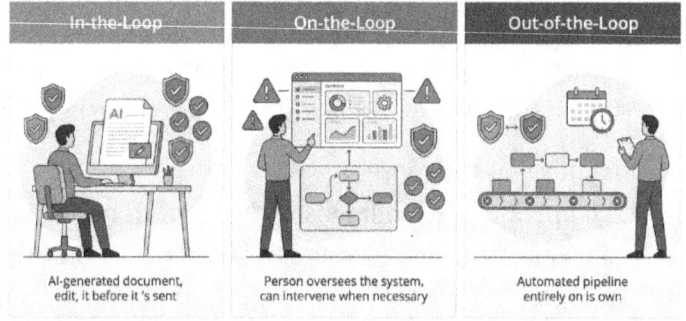

In-the-Loop	On-the-Loop	Out-of-the-Loop
AI-generated document, edit, it before it's sent	Person oversees the system, can intervene when necessary	Automated pipeline entirely on is own

---

9.7 Content risk management for generative AI

Content created by generative AI can introduce unique risks:

- Factual errors or hallucinations
  AI produces confident but incorrect statements that may mislead users.

- Tone and appropriateness
  AI-generated messaging may be technically correct but insensitive, unprofessional, or culturally inappropriate.

- Policy and legal misalignment
  AI may generate content that conflicts with internal policies or appears to commit the organization to positions it has not approved.

- Intellectual property and plagiarism
  AI may generate content too similar to proprietary or copyrighted materials, or incorporate protected concepts incorrectly.

To manage content risk:

- Define "use cases where generative AI is allowed" and where it is prohibited.

- Require human review for content in high-stakes contexts (e.g., legal responses, policy positions, sensitive communications).

- Provide style and policy guides for AI-generated content, and build them into prompts.

- Track examples of problematic outputs and use them to refine prompts, filters, and guidance.

Content risk management becomes part of your **Responsible AI** and **communications** governance, not just a technical concern.

9.8 Governance of autonomous workflows and agentic systems

Beyond content, autonomous and "agentic" systems can take actions, such as creating tickets, approving transactions, modifying settings, or orchestrating multiple tools. Governance must answer:

- What actions can the system take without human approval?

- What thresholds (value, risk) require human intervention?

- How are conflicts or ambiguous situations handled (e.g., competing signals)?

- How are actions logged and attributable to AI vs human users?

Control mechanisms should include:

- Action whitelists and blacklists
  Explicitly list allowed actions and forbid high-risk actions without human approval.

- Value and risk thresholds
  Allow small, low-risk actions to be handled automatically, but require review for larger or higher-impact actions.

- Simulation and sandbox testing
  Test autonomous behavior in controlled

environments before production, including edge cases.

- Strong logging and attribution
Ensure all AI-initiated actions are recorded with context so that incidents can be investigated and remediated.

As autonomy increases, your governance must become tighter and more systematic, not looser.

Action Boundaries for Autonomous Systems

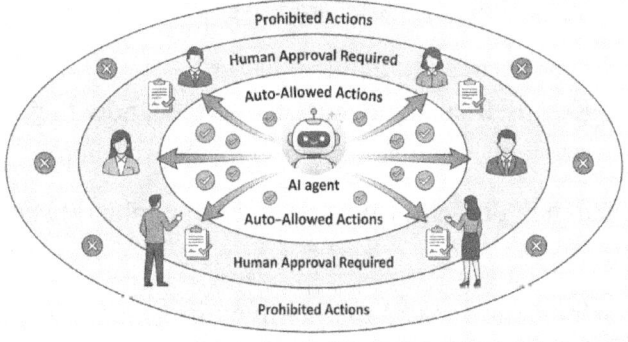

9.13 Documentation and transparency for generative and autonomous AI

Given their potential impact, generative and autonomous systems require strong documentation and transparency practices. At a minimum, for each such system, you should maintain:

- A clear system description
  What it does, where it is used, and for which user groups.

- Autonomy level and oversight pattern
  Whether it assists, augments, or acts, how humans oversee it.

- Guardrail configuration
  Input restrictions, prompts, filters, integrated tools, and any blocked actions.

- Risk assessment and approved scope
  Which risk tier applies, what assessment was done, and which use cases are in-scope vs out-of-scope.

- Monitoring and incident history
  Key metrics, incidents encountered, how they were resolved, and what improvements were made.

Transparency outward (to users and stakeholders) should at least include:

- Indication that AI is involved in generating content or taking actions where relevant.

- Clear paths to contact a human, request clarification, or challenge decisions.

This documentation should be part of your broader AI governance repository, not kept as separate notes within individual teams.

9.9 The manager's checklist: generative and autonomous AI

Use this checklist when confronted with a generative or autonomous AI proposal:

- What is the level of autonomy (assist, augment, act), and is that clearly documented?

- What are the highest-risk outcomes if the system behaves incorrectly or inappropriately?

- What human oversight mode are we using, and what exactly are humans expected to do?

- What guardrails (prompts, filters, blocked actions) are configured, and who manages them?

- How will we measure and monitor the system's behavior over time, including errors and complaints?

If you cannot answer these questions confidently, the system is not yet ready for deployment, especially in public-facing or high-stakes contexts.

9.10 AI Reality Check: "We'll just try it and see" is not a governance strategy

Generative and autonomous systems invite experimentation. The interfaces are intuitive, and early results can be impressive. This can tempt teams to say, "We'll just try it and see," especially for pilots. While experimentation is necessary, **ungoverned** experimentation with generative and autonomous AI can create real risk—even in pilots—because outputs can reach real people, and logs or training data may persist.

A more mature approach is **governed by experimentation**:

- Limit pilots to clearly defined contexts and user groups.

- Apply minimum guardrails and oversight even in tests.

- Treat pilot outputs with caution—label them, review them, and avoid irreversible actions.

- Document pilot scope, findings, and decisions about scaling or stopping.

As a manager, your role is not to shut down experimentation, but to shape it so that learning is safe, controlled, and aligned with your governance commitments.

9.11   How this chapter advances your governance journey

In this chapter, you adapted your governance approach to generative and autonomous AI, in which content creation, autonomy, and real-time interaction pose new risks and responsibilities. You learned to classify autonomy levels, design guardrails, define oversight patterns, manage content and action risks, and document systems with enough detail to support accountability and transparency. You also gained a practical checklist to evaluate proposals before they move from idea to reality.

In the next chapter, **Chapter 10 – People, Skills, and the Governance Culture**, we will shift focus to the human side: how to build the skills, incentives, and culture that make AI governance sustainable. Structures and processes are necessary, but without people who understand and believe in them, they will not deliver. The next chapter helps you ensure that managers, teams, and stakeholders are ready to live your governance framework, not just work around it.

10 People, Skills, and the Governance Culture

10.1 Opening vignette: "We have the framework, but not the follow-through."

An IT organization proudly rolled out its new AI governance framework. Policies were drafted, risk tiers defined, templates published, and a governance council chartered. For a few weeks, governance was a hot topic: leaders mentioned it in town halls, and Teams channels buzzed with links to the new documents. Yet, six months later, an internal review found that many AI initiatives still bypassed formal governance. Teams treated templates as optional. Risk assessments were copied from old projects. Frontline staff said they had never heard of the AI governance council. The framework existed—but only on paper. The missing ingredient was not more process; it was **people, skills, and culture**. This chapter is about closing that gap so your governance becomes lived practice, not just design.

10.2 Why Governance is a human system first

AI governance is often described in terms of structures, policies, and tools. Those are essential, but they all depend on people who understand them, apply them, challenge them, and sometimes work around them. If people see governance as irrelevant, burdensome, or disconnected from their reality, even the best-designed framework will fail. Conversely, if people believe governance protects mission, stakeholders, and their own work, they will use it as a guide rather than an obstacle.

As a manager, you must treat governance as a **human system**:

- Governance shapes how people make decisions under uncertainty.

- Culture determines whether people raise concerns or stay silent.

- Skills determine whether people can spot risks, interpret AI behavior, and apply controls.

This chapter focuses on building those human capabilities and attitudes so your governance can operate at scale and survive beyond any single leader or initiative.

10.3 The three pillars: mindset, skills, and incentives

To build a strong AI governance culture, you need to work across three pillars:

- Mindset
  How people think about AI, risk, responsibility, and their own role. Do they see governance as "someone else's job" or as part of their professional duty?

- Skills

  What people know and can do. Can managers ask the right governance questions? Can staff recognize when AI is stepping outside expected behavior?

- Incentives

  What people are rewarded or punished for. Are they encouraged to ship quickly at any cost, or to deliver value responsibly? Are they recognized for raising concerns and managing risk well?

You cannot rely on training alone. Mindset, skills, and incentives must reinforce each other. For example, teaching teams about AI risk without giving them time and recognition for using that knowledge will not change behavior.

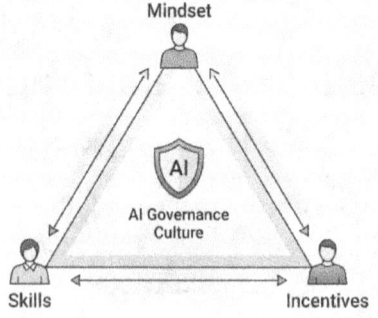

Three Pillars of Governance Culture

10.4 Defining governance roles for people, not just positions

Job titles vary, but governance roles are consistent. To build culture, people must know their role, regardless of their title. Core governance roles include:

- Sponsors and executives
  Set tone, approve frameworks, and back governance decisions—even when they slow down attractive projects.

- IT and data leaders
  Translate governance into technical practice, integrate controls into platforms and workflows, and support teams when trade-offs arise.

- Product and business owners
  Own AI use cases, balance value and risk, and ensure that governance is part of the product lifecycle, not a separate track.

- Risk, legal, and compliance partners
  Provide guidance, challenge assumptions, and help design controls and documentation that will withstand scrutiny.

- Frontline users and supervisors
  Interact with AI daily, notice issues first, and exercise human oversight when AI assists or automates decisions.

Clarifying these roles in simple terms—who does what, when, and why—helps people see themselves in the governance story and reduces the "not my job" tendency.

10.5 Building baseline AI governance literacy

You do not need everyone to become AI experts, but you do need **governance literacy** across key roles. At a minimum, people should understand:

- What AI is (and is not) in your context.

- Why AI governance matters for your mission, stakeholders, and personal accountability.

- The basic concepts: risk tiers, data integrity, fairness, explainability, and autonomy levels.

- Where AI governance processes live and how to use them: templates, review paths, incident reporting.

Tailor training by role:

- Executives: focus on risk appetite, trade-offs, and questions they should ask in reviews.

- Managers: focus on integrating governance into project planning, approvals, and oversight.

- Technical teams: focus on applying standards, documenting decisions, and collaborating with risk and business roles.

- Frontline staff: focus on interpreting AI outputs, escalation paths, and their right and duty to question AI.

The goal is not a one-time awareness session, but a **sustained baseline** that you refresh as your governance and AI landscape evolve.

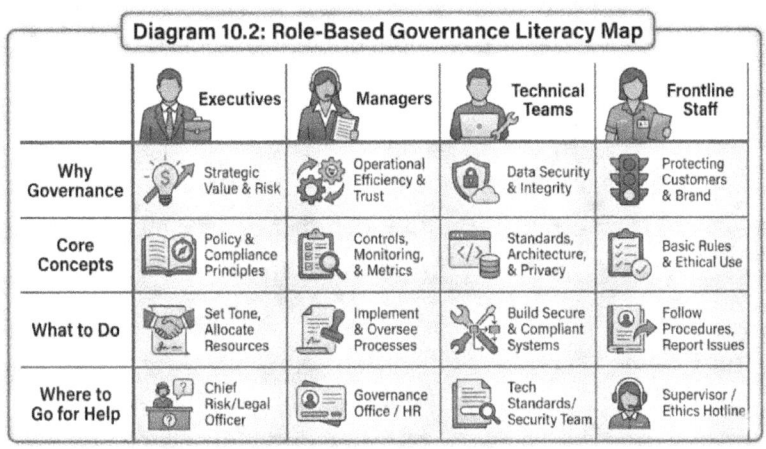

Diagram 10.2: Role-Based Governance Literacy Map

	Executives	Managers	Technical Teams	Frontline Staff
Why Governance	Strategic Value & Risk	Operational Efficiency & Trust	Data Security & Integrity	Protecting Customers & Brand
Core Concepts	Policy & Compliance Principles	Controls, Monitoring, & Metrics	Standards, Architecture, & Privacy	Basic Rules & Ethical Use
What to Do	Set Tone, Allocate Resources	Implement & Oversee Processes	Build Secure & Compliant Systems	Follow Procedures, Report Issues
Where to Go for Help	Chief Risk/Legal Officer	Governance Office / HR	Tech Standards/ Security Team	Supervisor / Ethics Hotline

10.8 Training for "governance moments."

The most impactful training is not generic; it is anchored in **governance moments**—the specific decision points where governance should influence behavior. Examples:

- When approving a new AI project or feature.

- When selecting a vendor or cloud AI service.

- When configuring a generative AI assistant for a team.

- When a frontline user sees an AI output that "doesn't feel right."

- When an incident or complaint related to AI arises.

Design short, scenario-based training around these moments:

- Present a realistic scenario.

- Ask, "What would you do?"

- Show how governance should guide the next steps (e.g., consult a risk template, escalate to a review board, log an incident).

This approach teaches people **how to act**, not just what to know. It also reinforces the idea that governance is present in everyday work, not just in formal meetings.

10.6   Manager behaviors that shape governance culture

Managers have disproportionate influence on culture. People watch what you do more than what you say. To build a governance-aware culture, managers should:

- Ask governance questions early and often
  For example, "Who is responsible for this AI system?", "What's the risk tier?", "How are we monitoring this?"

- Normalize raising concerns
  Respond positively when team members flag issues, even if the timing is inconvenient.

- Model transparency
  Admit uncertainty about AI risks and involve the

right partners, rather than pretending everything is fine.

- Respect governance decisions
  Support risk or ethics decisions even when they delay or limit a project, and explain why to your teams.

- Celebrate responsible decisions
  Recognize people who pause to reassess AI behavior, improve documentation, or proactively address risk—not just those who deliver features quickly.

These behaviors send a consistent message: governance is part of professional excellence, not optional bureaucracy.

Diagram 10.3: Manager Actions that Build Governance Culture

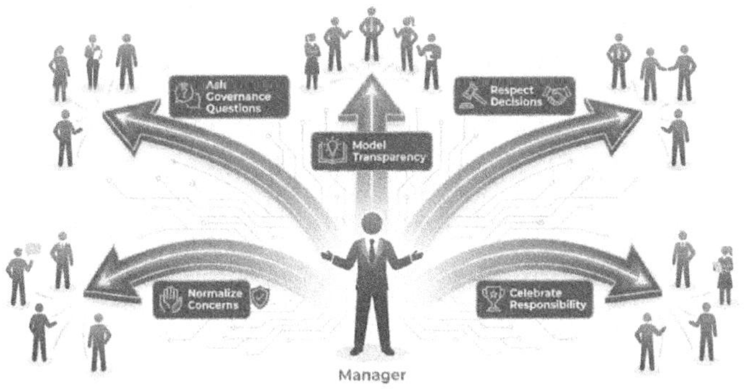

10.7 Handling resistance: common objections and responses

You will encounter resistance to governance. Typical objections include:

- "This slows us down."
  Response: Governance creates reusable patterns, reduces rework and incidents, and accelerates safe scaling.

- "This is someone else's job (risk, legal, IT, business)."
  Response: Governance is a shared responsibility; every role has specific duties, and you are accountable for your part.

- "We're just experimenting; this isn't real yet."
  Response: Even experiments can leak data, harm users, or set expectations. Governed experimentation is part of being responsible.

- "The vendor/cloud provider handles this."
  Response: You remain accountable for how AI is used in your context; vendors support but do not replace your governance.

Equip managers and champions with clear, concise responses to these objections. Over time, repetition will shift norms and expectations.

10.8 Building a network of AI governance champions

Formal structures are important, but informal networks often drive culture. You can accelerate adoption by creating a **network of AI governance champions** across departments and roles. These are people who:

- Understand the governance framework.

- Are interested in AI and responsible use.

- Are willing to help colleagues navigate processes and tools.

Support this network by:

- Providing extra training and early visibility into framework changes.

- Giving them a direct channel to governance leaders to share feedback and issues.

- Recognizing their contributions in performance reviews or public forums.

Champions act as multipliers, translating abstract frameworks into local practice and surfacing grassroots insights that central teams might miss.

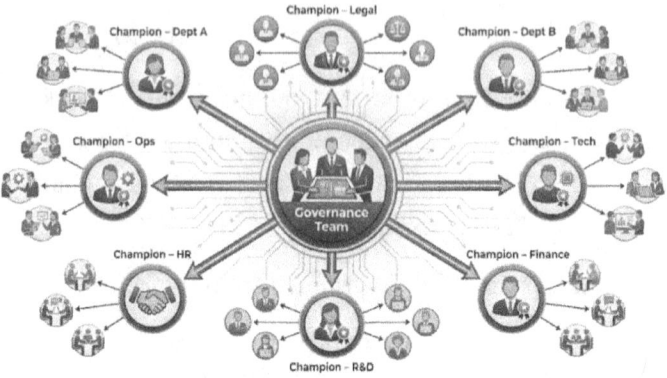

Diagram 10.4: AI Governance Champions Network

10.9 Aligning incentives, performance, and governance

Culture follows incentives. If people are rewarded solely for speed, volume, or cost-cutting, they will treat governance as a barrier. To align incentives with governance:

- Include responsible AI behaviors in performance goals
  For example, objectives related to documentation quality, risk management, or successful completion of governance gates.

- Recognize teams that handle incidents transparently and constructively
  Avoid punishing individuals for raising issues; focus on learning and improvement.

- Use metrics that include risk and quality, not just output
  For example, track AI incidents avoided or

154

resolved, governance compliance rates, and feedback from affected stakeholders.

- Incorporate governance into promotion and leadership criteria
Make it clear that leaders are expected to manage risk responsibly, not just deliver features.

These changes signal that governance is part of what "good" looks like, not a side task to be minimized.

10.10 The manager's checklist: building governance culture in your team

To make this chapter practical, here is a checklist you can use with your own team:

- Does everyone on my team know how AI affects our work and what governance processes apply?

- Have we identified our "governance moments" and built them into our routines (planning, reviews, stand-ups)?

- Do team members feel safe raising questions or concerns about AI behavior or risks?

- Have we identified at least one person on the team who can act as a governance champion?

- Do I, as a manager, regularly model the behaviors I want to see around governance (questions, transparency, respect for decisions)?

If the answer to several is "not yet," you have a clear agenda for cultural and skill-building work over the next quarter.

10.11 AI Reality Check: culture change is incremental, not instant

It is tempting to view culture change as a campaign—launch a new policy, hold a training, publish a message, and expect behavior to shift. In reality, **culture is shaped by repeated actions over time**. You will see pockets of early adoption and pockets of resistance. Some teams will embrace governance quickly; others will need more engagement, support, and sometimes pressure.

Do not wait for perfection before acting. Start small: embed governance into a few key workflows, support a few champions, and adjust incentives in a few teams. Celebrate progress publicly. Over time, these small changes accumulate into new norms. AI governance becomes "how we do things here," not an exception. As a manager, your persistence and consistency are more important than any single policy announcement.

10.12 How this chapter advances your governance journey

In this chapter, you shifted focus from frameworks and technology to the people who must live AI governance every day. You explored the three pillars of governance culture—mindset, skills, and incentives—and learned how to build role-based literacy, embed governance into "moments that matter," and shape manager behaviors that reinforce responsibility. You also saw how champions and aligned incentives can turn governance from a compliance narrative into a narrative of professional pride.

Next, in Chapter 11, you will turn to **Measurement, Maturity, and Continuous Improvement**—how to know where you stand, track progress, and evolve your AI governance over time. With people and culture now firmly in view, you will be ready to design metrics and maturity models that reflect not just processes on paper, but governance in practice.

11 Measurement, Maturity, and Continuous Improvement

11.1 Opening vignette: "Are we actually getting better, or just busier?"

An AI governance program had been running for over a year. New policies were in place, a council met monthly, and dozens of AI projects had passed through intake and review. The IT manager leading the effort felt constantly busy: answering questions, reviewing templates, and joining risk discussions. When the CIO asked, "So, how mature is our AI governance now, and what's improved since we started?" the room went quiet. There were anecdotes, but no shared picture. Some teams said governance was working; others claimed it was slowing them down. Risk and compliance partners felt more involved, but were still unsure whether key AI risks were being systematically managed. The missing piece was clear: the organization had implemented governance **activities**, but had not defined how to **measure governance maturity or outcomes**. This chapter is about avoiding that situation and building a measurement-and-improvement loop around your AI governance.

11.2 Why measuring AI governance matters

If you cannot measure your AI governance, you cannot convincingly answer basic questions:

- Are we reducing AI-related risk over time?

- Are we enabling responsible AI adoption efficiently, or creating friction without value?

- Where should we focus our limited improvement efforts next?

Measurement is not about turning governance into a scorecard for its own sake. It is about:

- Providing leadership with visibility into progress and gaps.

- Giving teams feedback so they can improve.

- Prioritizing where to invest in processes, tools, and skills.

Without measurement, governance risks become static: frozen at their initial design, disconnected from real-world outcomes. With measurement, governance becomes a **learning system**—able to adapt as AI, regulations, and organizational needs evolve.

11.3 The governance maturity lens

A practical way to think about AI governance progress is through a **maturity lens**. Rather than asking "Do we have governance or not?", you ask "At what level of maturity are we across key dimensions?" A simple, manager-ready maturity scale might be:

- Level 1 – Ad hoc
  AI governance is informal, based on individual

judgment. Some good practices exist, but they are inconsistent and undocumented.

- Level 2 – Emerging
  Basic policies, roles, and processes are defined. Governance applies to some AI projects, but coverage is incomplete, and understanding varies.

- Level 3 – Established
  AI governance is consistently applied to the most relevant AI systems. Roles are clear, templates are used, and monitoring is in place, though not yet optimized.

- Level 4 – Optimized
  Governance processes are efficient, integrated into workflows, and supported by tools. Metrics are tracked and used to improve both governance and AI outcomes.

Not every organization needs to reach Level 4 in all areas immediately. The key is to know where you are now and where you intend to be in the near term, given your risk profile and ambitions.

AI Governance Maturity Staircase

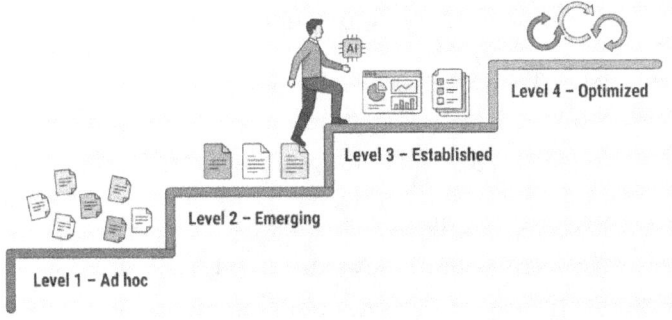

11.4 Key dimensions of AI governance maturity

To make maturity actionable, break it into dimensions you can evaluate separately. Typical dimensions include:

- Strategy and scope
  How clearly AI governance objectives, scope, and risk appetite are defined and communicated.

- Structures and roles
  How well are roles, committees, and decision rights for AI governance defined and functioning?

- Policies and processes
  How complete and usable are your AI-related policies, standards, and lifecycle processes?

- Data and model governance
  How robust are your data integrity, documentation, and model lifecycle practices for AI?

- Risk and compliance integration
  How well AI risks are integrated into enterprise risk management and compliance activities.

- Ethics and transparency practices
  How consistently are ethics and transparency applied in AI design, deployment, and communication?

- People and culture
  How widely are AI governance literacy, behaviors, and incentives embedded in the organization?

- Tooling and automation
  How much governance is supported by tools (e.g., registries, monitoring, workflow systems) versus manual effort?

For each dimension, you can assess your current level and desired target using the maturity scale. This becomes your roadmap.

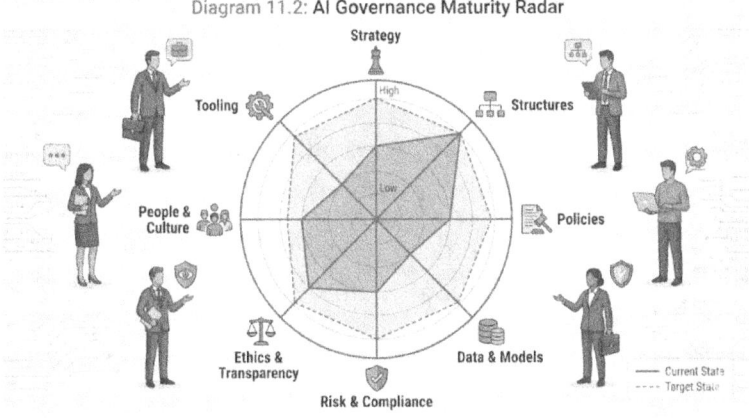

Diagram 11.2: AI Governance Maturity Radar

11.5   Defining meaningful metrics: outputs vs outcomes

Metrics for AI governance fall into two broad categories:

- Output metrics (activity and coverage) These measure whether governance processes are being followed. Examples:

  - Number and percentage of AI projects that went through formal intake and risk assessment.

  - Number of models recorded in an AI registry with complete documentation.

  - Time taken from AI proposal to approval for each risk tier.

- Outcome metrics (effectiveness and impact) These measures assess whether governance influences real-world outcomes. Examples:

163

o Number of AI incidents (e.g., fairness complaints, unexpected behavior) and their severity.

o Trend in AI-related audit findings or regulatory questions.

o Stakeholder trust indicators (e.g., user satisfaction with AI-supported services, internal survey results on confidence in AI use).

You need both: output metrics show that governance is active; outcome metrics show that governance is **effective**. Start simple, with a handful of metrics you can realistically track and discuss, then expand as your capability grows.

11.6 Selecting a small, high-value metric set

Rather than attempting to measure everything, choose a small set of high-value metrics aligned with your current priorities. For example, an organization early in its governance journey might focus on:

- Percentage of AI use cases captured in the central registry.

- Percentage of high-risk AI use cases with completed risk assessments and approvals.

- Number of AI incidents or complaints logged per quarter.

A more mature organization might add:

- Average time to resolve AI incidents.

- Percentage of AI systems with up-to-date monitoring dashboards and periodic reviews.

- Number of improvements made as a result of governance (e.g., models retrained, policies updated).

Review your metrics periodically to ensure they remain relevant and do not incentivize superficial compliance (e.g., filling templates without meaningful review).

11.7   The governance measurement loop

Measurement only helps if it feeds into a **loop** of review and improvement. A simple, repeatable loop:

1. Collect
   Gather data on your chosen metrics regularly (monthly or quarterly).

2. Analyze
   Look for patterns: where are you improving, where are you stuck, where are incidents clustering?

3. Discuss
   Bring metrics to governance forums (e.g., AI council, risk committees) and management meetings for collective interpretation.

4. Decide
   Identify specific improvement actions (e.g., refine a template, add a training, adjust risk tiers, invest in tooling).

5. Act and document
   Implement changes and capture them as part of governance evolution.

6. Reassess
   See how metrics respond to those changes in subsequent cycles.

This loop turns metrics into **management tools**, not just reports.

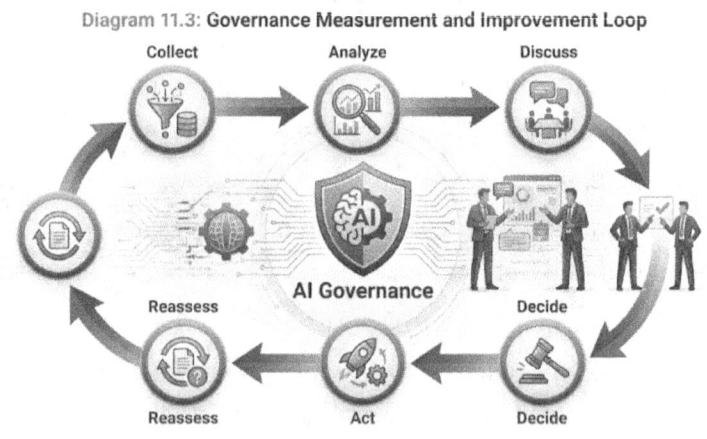

Diagram 11.3: **Governance Measurement and Improvement Loop**

11.8 Using maturity assessments as a leadership tool

Periodic maturity assessments—formal or informal—can be powerful leadership tools. They:

- Provide a shared language for where you are and where you're going.

- Help align expectations between IT, business, and risk functions.

- Support budgeting and prioritization decisions (e.g., investment in tooling or training).

A practical approach:

- Develop a short self-assessment survey based on your maturity dimensions and levels.

- Ask key stakeholders (IT, business, risk, compliance, data teams) to rate the organization independently.

- Facilitate a session to discuss differences, agree on a consensus view, and identify 3–5 priority improvement areas for the next year.

Repeat the assessment annually. Over time, you will see which areas advance, which stall, and why. This turns maturity from a one-time exercise into a **strategic planning input**.

11.9   Linking governance metrics to AI and business outcomes

To keep governance meaningful, connect your metrics to outcomes that matter to leadership and mission owners. For example:

- Reduced AI incident frequency or severity
  Demonstrates improved risk management and fewer crises.

- Faster time-to-production for governed AI projects
  Shows that governance streamlines responsible deployment rather than blocking it.

- Improved stakeholder trust indicators
  Internal: more teams choosing to use governed AI tools.
  External: higher satisfaction with AI-enabled services or fewer public complaints.

When you can say, "Because we improved governance in these ways, we saw these concrete benefits," governance moves from cost to **value-generating capability**.

11.10   The manager's checklist: measuring what matters

Use this checklist to start or refine your governance measurement approach:

- Have we defined a small set of clear AI governance metrics (both output and outcome)?

- Do we have a simple maturity model and dimensions, and have we assessed ourselves at least once?

- Are metrics and maturity results regularly reviewed in governance and leadership forums?

- Do we use measurement insights to make explicit decisions about where to improve governance?

- Can we tell a credible story about how AI governance has improved risk management or delivery over the last year?

If you cannot yet answer "yes" to most of these, your measurement strategy is a high-impact area for development.

11.11 AI Reality Check: imperfect data is better than no data

Many organizations hesitate to measure governance because metrics or maturity assessments feel imprecise. They worry about disagreements between teams or incomplete data. This can lead to inaction—operating "by feel" rather than by evidence.

The reality is that **imperfect but honest data** is far better than no data. An approximate count of AI systems is better than none; rough risk tier coverage estimates are better than guesses. Over time, your measurement will improve, but only if you start. Treat early metrics and maturity assessments as hypotheses, not final judgments. The value is in the conversation and improvement, not in the illusion of precision.

11.12 How this chapter advances your governance journey

In this chapter, you learned how to move from "doing governance" to **measuring and improving governance**. You explored maturity levels and dimensions, defined meaningful metrics, and saw how to build a measurement and improvement loop that keeps your governance responsive and aligned with outcomes. You also gained practical checklists and tools to integrate measurement into leadership discussions and planning.

The final chapter, **Chapter 12 – The Governance Playbook for AI-Modern IT**, will bring everything together. It will synthesize the frameworks, practices, and cultural elements from the previous chapters into a concise, actionable playbook to guide your organization through AI modernization with confidence and responsibility.

12 The Governance Playbook for AI-Modern IT

12.1 Opening vignette: "We need a playbook, not another policy."

The CIO sat with her direct reports after a turbulent year of AI experimentation. Some initiatives had delivered real value; others had created headaches. A generative AI assistant improved internal documentation but raised questions about data use. A risk-scoring model needed to be pulled back after fairness concerns. Governance structures were emerging, but teams still asked the same question: "What do we actually do, step by step, when someone brings us a new AI idea?" Leadership did not want another long policy document. They wanted a **playbook**—a concise, practical guide that managers could follow from idea to production, combining everything the organization had learned about AI governance. This chapter is that playbook: a structured, manager-ready sequence you can adapt to your own organization.

12.2 What this playbook is (and is not)

This playbook is a **practical guide** for IT managers and their partners to move AI initiatives from idea to safe, mission-aligned operation. It brings together core elements from previous chapters—strategy, operating model, risk, ethics, data, cloud, people, and measurement—into a single, repeatable flow.

It is not:

- A replacement for your detailed policies, standards, or legal advice.

- A one-size-fits-all blueprint that ignores your specific context.

- A checklist to be followed mechanically without judgment.

Instead, it is a **scaffold**: a structured path that you can customize, expand, or simplify based on your sector, risk profile, and maturity.

12.3   The AI governance playbook at a glance

The playbook can be summarized in eight stages:

1. Discover and frame

2. Assess and classify

3. Design and align

4. Build and document

5. Test and challenge

6. Deploy and integrate

7. Monitor and respond

8. Learn and improve

Each stage includes key questions, actions, and artifacts. Together, they provide the "Monday morning" roadmap for managers leading AI modernization.

Eight-Stage AI Governance Playbook

Optimized for executive presentation slide

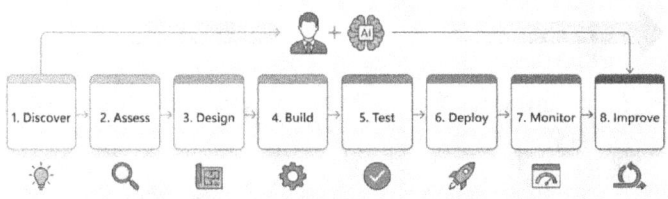

12.4 Stage 1 – Discover and frame

Goal: Ensure every AI idea starts with clarity of purpose, context, and stakeholders.

Key questions:

- What problem are we trying to solve, and why does it matter?

- Who will be affected by this AI system (internal and external)?

- How does this idea support our mission and strategy?

Manager actions:

- Ask for a short, plain-language description of the AI idea and expected benefits.

- Identify primary stakeholders and any groups that may be at risk.

- Check whether similar initiatives exist to avoid duplication and leverage existing work.

Artifacts:

- One-page use case summary: purpose, value, stakeholders, initial thoughts on data and AI type.

This stage avoids "AI for AI's sake" and sets the tone that governance begins with understanding people and mission.

12.5   Stage 2 – Assess and classify

Goal: Determine the **risk tier** and complexity to ensure governance is proportional.

Key questions:

- What kinds of decisions or actions will this AI system influence?

- What could go wrong, and how severe would the consequences be?

- What data will be used, and how sensitive is it?

Manager actions:

- Use your AI risk tiering model (e.g., Tiers 1–4) to classify the initiative.

- Identify the level of autonomy (assist, augment, act).

- Decide which governance path applies (lightweight vs full review).

Artifacts:

- Initial AI risk classification form: risk tier, autonomy level, affected domains.

This stage ensures that high-impact initiatives receive stronger governance and that low-risk experiments are not overburdened.

Diagram 12.2: **From Idea to Risk Tier**

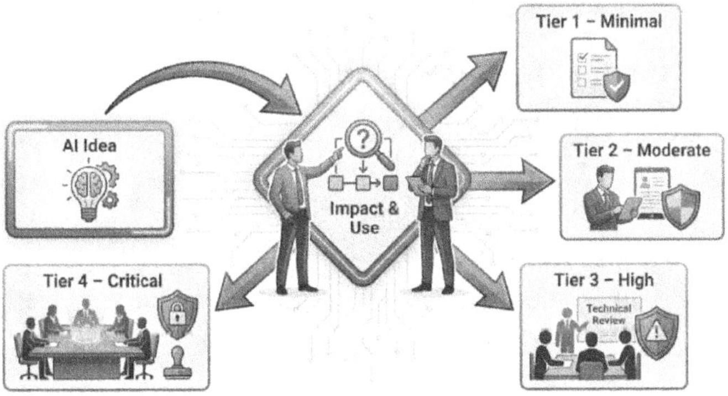

12.6   Stage 3 – Design and align

Goal: Align design choices with governance principles, operating model, and ethical expectations **before** building.

Key questions:

- How will this AI system fit into existing workflows and systems?

- What ethical, fairness, and transparency considerations are most important here?

- Who owns this use case, data, and model, and where will it run (on-prem, cloud, hybrid)?

Manager actions:

- Identify product/usage owner, data owner, and technical leads.

- Decide on the autonomy level and human oversight pattern (in-the-loop, on-the-loop, etc.).

- Engage risk, legal, compliance, and data governance partners as appropriate for the risk tier.

- Confirm where the system will be hosted and which providers are involved.

Artifacts:

- Design and governance concept document: roles, data sources, oversight pattern, ethical considerations, and hosting model.

This stage connects the AI idea to your **AI operating model** and governance framework, preventing technical design from drifting away from policy and risk appetite.

12.7 Stage 4 – Build and document

Goal: Build the AI solution within approved guardrails and capture the key design and data decisions.

Key questions:

- Are we using approved platforms, tools, and data sources?

- Are data integrity and lineage being maintained?

- Are we documenting choices so we can explain them later?

Manager actions:

- Ensure teams use sanctioned data, platforms, and cloud services.

- Require documentation of data sources, preprocessing, and feature engineering.

- Make sure model choices (or vendor AI services) are documented with rationale and limitations.

Artifacts:

- Model and data documentation ("model card" and data summary).

- Configuration records for prompts, filters, and integration settings (for generative and cloud systems).

This stage ensures technical work is done within the **governed rails** and that key information is not lost in developers' heads or ad hoc notes.

12.8   Stage 5 – Test and challenge

Goal: Validate performance, fairness, robustness, and security before deployment—and challenge assumptions.

Key questions:

- Does the system perform well enough in realistic scenarios?

- Are there fairness, explainability, or safety concerns in key groups or edge cases?

- How does the system behave under misuse or adversarial inputs (where relevant)?

Manager actions:

- Ensure standard testing suites include: functional tests, performance metrics, fairness checks, and explainability assessments.

- Involve cross-functional reviewers (e.g., risk, data governance, ethics champions) for higher-risk systems.

- Encourage "red-teaming" for critical or generative/autonomous use cases—actively trying to break or misuse the system.

Artifacts:

- Test and validation report: performance metrics, subgroup analyses, explanation examples, robustness tests, and mitigation steps.

This is where you move from "it seems to work" to "we have evidence it behaves acceptably under our governance standards."

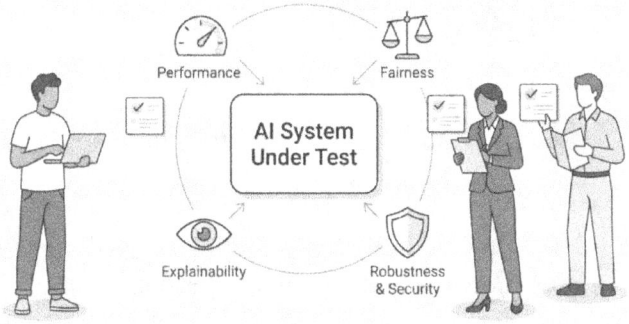

Test & Challenge

12.9   Stage 6 – Deploy and integrate

Goal: Deploy the AI system into production with a clearly defined scope, controls, and responsibilities.

Key questions:

- What exactly are we approving this system to do (and not do)?

- How will we handle rollout: pilot, phased, or full deployment?

- Who is accountable for operations, monitoring, and user support?

Manager actions:

- Obtain approvals appropriate to the risk tier (e.g., project owner, governance council, risk, or legal sign-off).

- Define deployment scope and limitations (e.g., specific user groups, transaction thresholds).

- Ensure integration with existing workflows includes user instructions, override mechanisms, and escalation paths.

Artifacts:

- Deployment approval record, including risk tier, scope, guardrails, and conditions.

- Operational runbook: contacts, alerts, incident procedures, and maintenance responsibilities.

This stage is where governance decisions become explicit: who said "yes," under what conditions, and with what expectations for safe operation.

12.10 Stage 7 – Monitor and respond

Goal: Continuously monitor AI behavior, detect issues early, and respond effectively.

Key questions:

- How will we know if the system is drifting, failing, or causing unintended harm?

- What metrics will we track, and who will review them?

- What happens when users or stakeholders raise concerns?

Manager actions:

- Ensure monitoring dashboards track key metrics: performance, usage, error rates, fairness indicators, override rates, and relevant incidents.

- Set thresholds for alerts and define who receives them and how quickly they must respond.

- Establish a clear process for logging and investigating AI-related incidents or complaints.

Artifacts:

- Monitoring dashboards and threshold definitions.

- Incident log with resolution notes and follow-up actions.

Monitoring closes the loop between design assumptions and real-world behavior, transforming governance from a one-time event into a continuous safety net.

12.11 Stage 8 – Learn and improve

Goal: Use real-world data, incidents, and feedback to improve both AI systems and governance itself.

Key questions:

- What have we learned from this AI system's operation so far?

- Do we need to retrain, recalibrate, or retire the model?

- What governance processes or standards should we adjust based on our experience?

Manager actions:

- Conduct periodic reviews for higher-risk AI systems (e.g., quarterly or semi-annually).

- Summarize key findings: what worked, what failed, what surprised us.

- Propose and implement adjustments to models, workflows, policies, or training.

Artifacts:

- Periodic review reports for significant AI systems.

- Governance change log documenting improvements to processes, templates, or standards.

This stage turns every AI system into a **source of governance learning**, feeding improvement into both technical and organizational practices.

Diagram 12.4: Playbook in a Continuous Loop

12.12 The manager's condensed playbook checklist

To make this chapter immediately usable, here is a condensed checklist you can keep in front of you whenever an AI initiative appears:

1. Discover and frame

   o Do we understand the problem, value, and stakeholders?

2. Assess and classify

   o Have we assigned a risk tier and autonomy level?

3. Design and align

   o Do we know the owners, data, oversight pattern, and hosting model?

4. Build and document

   o Are we using approved tools and capturing key data/model decisions?

5. Test and challenge

   o Have we tested performance, fairness, explainability, and robustness, and challenged assumptions?

6. Deploy and integrate

   o Has the right authority approved deployment with clear scope and guardrails?

7. Monitor and respond

    o Are monitoring and incident processes in place and understood?

8. Learn and improve

    o Are we reviewing and improving both the AI system and our governance based on experience?

If you cannot answer "yes" to each question for a given AI system, you have identified where to focus next.

12.13 AI Reality Check: your playbook will evolve

No playbook is perfect on day one. Technologies advance, regulations change, and your organization's AI portfolio will grow and diversify. The true mark of a mature AI governance function is not a static playbook, but a **living one**—updated as you learn, simplified where it's too heavy, and strengthened where incidents show gaps.

Treat this chapter's playbook as your first integrated draft. Use it with a few AI initiatives, observe where it works and where it creates friction, and adjust. Over time, your organization will develop its own "house style" of AI governance—rooted in the principles and structures from this book, but tailored to your mission, constraints, and people. Your role as a serious AI manager is to keep that playbook both **disciplined** and **adaptable**, so governance remains a source of clarity and confidence in an AI-modern IT landscape.

13 AI Portfolio Management and Prioritization

13.1   Opening vignette: "Too many pilots, not enough progress."

The IT manager reviewed a slide titled "AI Initiatives – Current State." The list was long: a chatbot pilot in customer service, a predictive maintenance proof-of-concept in operations, a fraud detection model in finance, a generative assistant for policy drafting, and several "experimental" models running in sandboxes. Each team was excited about its project. Individually, many looked promising. Collectively, they formed a chaotic picture: duplicated efforts, inconsistent governance, unclear ownership, and no clear line of sight to enterprise strategy. When the CIO asked, "Which of these should we scale, which should we stop, and where are the biggest risks?" there was no structured answer. The problem was not a lack of ideas—it was the absence of **AI portfolio management**. This chapter shows you how to move from a scattered set of AI experiments to a governed AI portfolio that advances your mission.

13.2 Why AI portfolio management matters for governance

AI governance at the individual project level is necessary, but not sufficient. You also need a portfolio lens that lets you see:

- Which AI initiatives exist and where they sit in their lifecycle.

- How resources, risks, and value are distributed across domains.

- Whether you are unintentionally overexposed to particular risks or underinvesting in foundational capabilities.

Without portfolio management:

- High-risk experiments can slip through because they look small in isolation.

- Low-value or duplicative projects can consume scarce talent and infrastructure.

- Leadership cannot easily answer basic questions about AI strategy, value, and exposure.

A governed AI portfolio turns AI from a collection of local experiments into a coordinated enterprise capability. It allows you to **prioritize**, **sequence**, and **shape** AI investments in line with your overall governance framework and operating model.

13.3   What is an AI portfolio (in practical terms)?

An AI portfolio is a structured view of all significant AI initiatives in your organization, past and present. It includes not just "big projects," but also pilots, embedded vendor AI features, and critical models already in production. For each initiative, the portfolio should capture:

- Purpose and business/mission area.

- Risk tier and autonomy level.

- Current lifecycle stage (idea, pilot, production, decommissioned).

- Owner (business/product) and technical lead.

- Data domains used.

- Value expectations (qualitative or quantitative).

This portfolio view helps you make management decisions:

- Where to double down.

- Where to pause or retire.

- Where to invest in shared capabilities (data, platforms, skills).

The AI portfolio is not a static document. It must be maintained and reviewed regularly, becoming a living record of how AI evolves in your environment.

13.4 Diagram 13.1: "From scattered projects to a governed AI portfolio."

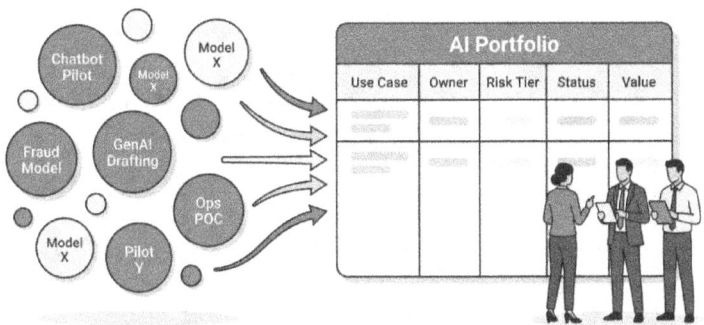

From Scatter to Portfolio

13.4 Core dimensions of AI portfolio management

To manage an AI portfolio systematically, define consistent dimensions for each initiative. At minimum:

- Strategic alignment
  How strongly the initiative supports key strategic or mission objectives.

- Value potential
  The expected impact on outcomes (e.g., efficiency, quality, citizen satisfaction, revenue, cost avoidance).

- Risk and autonomy
  The risk tier and level of autonomy are defined in earlier chapters.

- Data readiness
  How ready and governed the required data is (e.g.,

high-quality and accessible vs fragmented and untrusted).

- Feasibility and complexity
  Technical and organizational complexity, including dependencies on other systems or teams.

- Maturity and scalability
  Stage of development (idea, pilot, production) and potential for reuse or scaling to other units.

These dimensions create a **common language** for comparing and prioritizing AI initiatives across domains and teams.

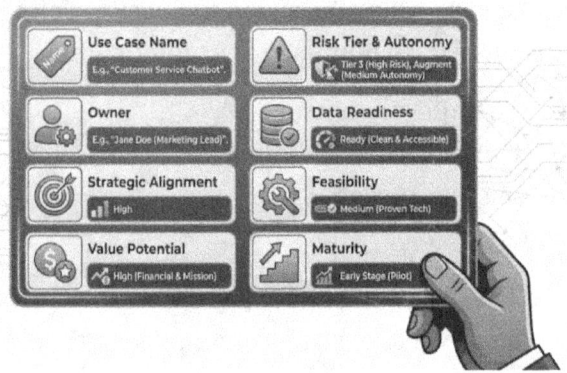

Diagram 13.2: **Portfolio Dimension Card**

13.5 From list to insight: visualizing the AI portfolio

Once you capture basic information for each initiative, you can visualize the portfolio to support better decisions. Practical visualizations include:

- Value–risk grid
  Plot initiatives by value potential (vertical axis) and risk/complexity (horizontal axis). Quick wins appear as high-value/low-risk; strategic bets as high-value/high-risk.

- Lifecycle pipeline
  Show counts of initiatives in each stage: discovery, pilot, production, and decommissioned. This reveals bottlenecks (e.g., many pilots, few scaled systems).

- Domain distribution
  Show where AI initiatives are concentrated (e.g., customer service, operations, finance). This highlights over- or under-invested areas.

- Risk exposure view
  Show the number of initiatives by risk tier and domain, indicating where governance attention should focus.

These visualizations can become standard artifacts for governance and leadership forums, replacing ad hoc slides with a coherent picture.

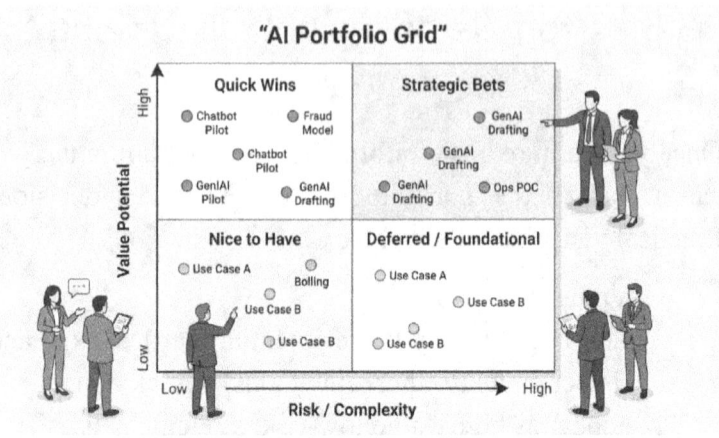

"AI Portfolio Grid"

13.6 Prioritization: choosing where to invest, pause, or stop

Portfolio management alone does not decide; you still must **prioritize**. A practical approach:

- Elevate initiatives that:

    o Have strong strategic alignment.

    o Show clear value potential.

    o Sit in manageable risk tiers, or high-risk but critical domains with strong governance readiness.

- Deprioritize or pause initiatives that:

    o Have weak or unclear value.

    o Lack data readiness or rely on fragile data sources.

- o Fall into high-risk categories without a credible governance path.

- Retire initiatives that:

  - o Have been piloted repeatedly without compelling value.

  - o Duplicate capabilities are better provided elsewhere.

  - o Create a governance burden disproportionate to their benefit.

These decisions should be made in cross-functional forums (e.g., your AI or tech governance council) using consistent criteria, not just loudest voices or latest trends.

13.7 Aligning portfolio management with your AI operating model

Portfolio management must connect to your AI operating model (from Chapter 3). Specifically:

- Intake and demand
  All new AI ideas should enter through recognized channels that feed the portfolio, not bypass it.

- Capacity planning
  The portfolio informs how many initiatives your central and federated teams can realistically support, given governance, data, and platform constraints.

- Shared capabilities
  The portfolio reveals patterns (e.g., many initiatives requiring similar data or tools), which guide investment in shared platforms, MLOps, or governance automation.

- Role clarity
  Portfolio decisions must involve product owners, IT, data, and risk leads, mirroring the roles defined in the operating model.

The portfolio becomes the **bridge** between strategy, operating model, and project execution.

13.8   Governance council as portfolio steward

Your AI or technology governance council should act as the steward of the AI portfolio. Its responsibilities include:

- Reviewing the portfolio periodically (e.g., quarterly).

- Approving or reprioritizing initiatives based on value, risk, and readiness.

- Ensuring that high-risk or cross-cutting initiatives receive appropriate oversight.

- Identifying where governance, data, or platform investments are needed to unlock more value.

This moves the council from being purely a review body to a **strategic portfolio body**, influencing where the organization places its AI bets.

13.12 The Manager's Playbook: running a quarterly AI portfolio review

To operationalize portfolio governance, managers can structure a quarterly review as follows:

- Step 1: Refresh portfolio data
  Ensure owners update status, metrics, and key changes for each initiative.

- Step 2: Present summary views
  Show the portfolio grid, domain distribution, lifecycle pipeline, and risk exposure views.

- Step 3: Discuss hot spots and gaps
  Where are we overconcentrated? Where is there latent value but low activity? Are certain risk tiers under-governed?

- Step 4: Make decisions
  Confirm which initiatives to accelerate, pause, or retire. Identify enabling investments (data, tools, skills).

- Step 5: Communicate outcomes
  Share decisions and rationales with initiative owners and key stakeholders, linking them to governance and strategy.

Over time, this cadence creates predictability and reinforces that AI initiatives must justify their place in a governed portfolio, not just exist as independent experiments.

13.9 The manager's checklist: AI portfolio basics

To close this chapter, use this checklist to assess your current portfolio readiness:

- Do we maintain a single, up-to-date inventory of significant AI initiatives?

- Does each entry have basic information: purpose, owner, risk tier, status, and domain?

- Do leadership and governance forums regularly review portfolio views, not just individual projects?

- Are prioritization decisions documented and explained in terms of value, risk, and readiness?

- Do we use portfolio insights to adjust our AI operating model, governance focus, and enabling investments?

If the answer to several of these is "not yet," establishing an AI portfolio practice is one of the highest-leverage steps you can take to mature your AI governance.

14 What is Change Management and Adoption for AI-Governed Systems

14.1 Opening vignette: "The AI was good, the rollout wasn't."

A large service organization invested heavily in an AI-powered case triage system. Governance was thorough: risk assessments were completed, data quality was vetted, fairness was tested, and approvals were obtained. Technically, the launch was a success—the system integrated cleanly with existing tools and performed well in simulations. Yet three months after deployment, usage statistics told a different story. Many frontline staff were bypassing AI recommendations, reverting to manual queues. Some supervisors instructed teams to "use it only if you have time," and others quietly turned off certain features, worried they would be blamed for AI mistakes. The AI system itself was sound; the **change management and adoption strategy** was not. Governance had focused on the system, but not enough on the people expected to live with it. This chapter focuses on how to bridge that gap.

14.2   Why change management is central to AI governance

AI governance often concentrates on approvals, risk assessments, and monitoring. Those are crucial, but if people do not understand, trust, or use AI systems as intended, all that work fails to deliver value—and can even create new risks:

- Under-adoption
  People ignore AI and stick to legacy processes. Value is lost, and leadership misreads "AI is in production" as "AI is working."

- Misuse
  People over-trust AI outputs or use tools in ways governance never anticipated, creating ethical, legal, or operational exposure.

- Workarounds
  Teams build informal shortcuts around AI constraints, undermining controls and making monitoring data misleading.

Change management connects the **governance framework** you've built to real behaviors in daily workflows. It ensures that people understand why AI is used, what governance means for them, and how they should act when AI and reality diverge.

14.3 The human impact of AI-governed change

Any AI deployment changes how people work. With governance in place, it also changes:

- Who is accountable for decisions?
- How much discretion individuals have versus the system.
- What is expected when AI outputs look questionable?

Common human concerns include:

- "Will this replace my judgment or my job?"
- "If I follow the AI and it's wrong, who gets blamed?"
- "If I ignore the AI, am I breaking the rules?"
- "Do the people designing this even understand my work?"

If these questions are unanswered, resistance—active or passive—is rational. Effective change management for AI-governed systems anticipates these concerns and addresses them explicitly, linking **governance** to **psychological safety** and **professional pride** instead of fear.

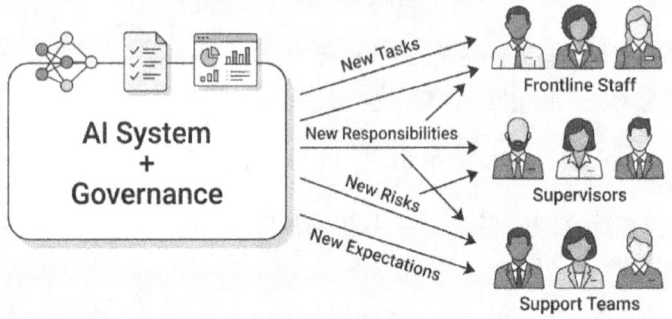

14.4 Mapping stakeholders and their change journeys

Before deploying an AI-governed system, identify and map key stakeholder groups:

- Frontline users
  People who will interact directly with AI outputs or tools (e.g., agents, analysts, clinicians, inspectors).

- Supervisors and middle managers
  Those who oversee performance, coach users, and interpret governance expectations.

- Support and operations staff
  Teams that handle system issues, data corrections, or escalation.

- Risk, legal, and compliance partners
  Stakeholders who must understand how AI is used to fulfill their oversight roles.

- External stakeholders (where applicable)
  Citizens, customers, or partners affected by AI-supported decisions.

For each group, consider:

- What changes in their daily work?

- What benefits and risks do they perceive?

- What questions about accountability and fairness will they have?

This mapping informs your communication, training, and support plans, ensuring they are relevant and targeted rather than generic.

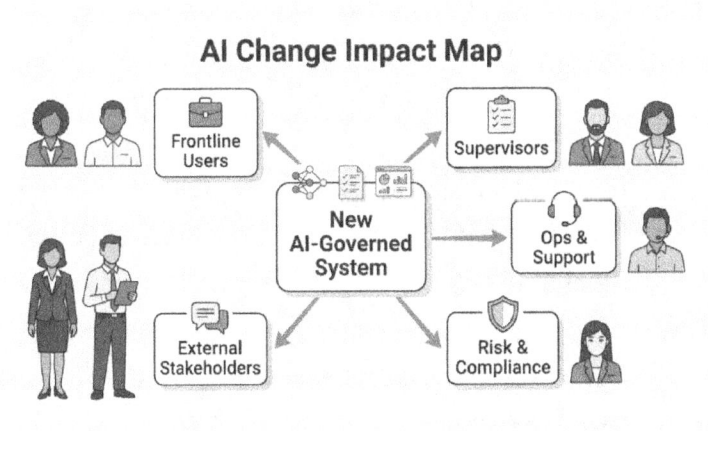

14.5 Designing change for governed AI: five design questions

For each AI rollout, managers should systematically address five change-design questions:

1. What is changing and why?
   Describe the change in plain language: what people will do differently, why AI is being introduced, and how it supports mission and governance.

2. What will stay the same?
   Clarify which parts of the workflow, roles, or responsibilities remain unchanged to reduce anxiety and ambiguity.

3. What does governance mean for each role?
   Explain what risk tier, oversight pattern (in-the-loop, on-the-loop), and policies mean for daily decisions.

4. How will we support people through the change?
   Identify training, job aids, help channels, and time for learning.

5. How will we listen and adapt?
   Define feedback mechanisms and how they connect back to governance and improvement.

Answering these questions concretely for each major AI deployment provides a blueprint for responsible change.

14.6 Training that reflects governance, not just features

Traditional system training focuses on how to click buttons and navigate screens. For AI-governed systems, training must also cover:

- The AI's role in decisions
  Where AI is assisting vs acting; what level of autonomy applies.

- Oversight expectations
  When users must review or challenge AI outputs, examples of acceptable and unacceptable reliance are provided.

- Escalation and incident pathways
  How to report issues, suspected bias, or unexpected behavior; who will respond.

- Accountability clarity
  Who is ultimately accountable for decisions, and how does governance protect staff when they follow agreed-upon practices?

Training should use realistic scenarios where AI suggests something plausible but wrong, where governance requires a different choice, and where users must practice acting within their oversight responsibilities.

Training Layers for AI-Governed Systems

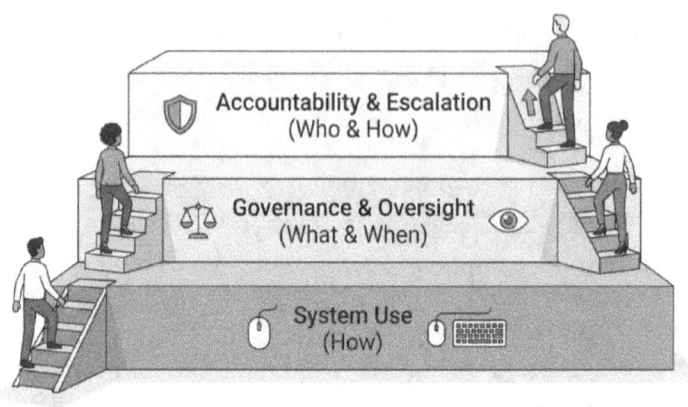

14.7 Shaping trust: avoiding both blind faith and deep suspicion

Adoption risk often takes two extreme forms:

- Blind faith: "The AI is smarter, so I should just follow it."

- Deep suspicion: "This is untrustworthy; I'll ignore it."

Governance-aware change management aims for **informed trust**:

- Users understand what AI is good at and where it can fail.

- They know which outputs require careful review versus routine acceptance.

- They see evidence that fairness, risk, and data integrity have been considered.

Managers can shape trust through:

- Transparent communication about testing and limitations.

- Sharing examples of both AI success and AI mistakes caught by oversight.

- Reinforcing that questioning AI is encouraged and expected in certain scenarios.

Trust grows when people see that governance is real, responsive, and aligned with their professional judgment—not in conflict with it.

14.8   Supporting supervisors: the bridge between policy and practice

Supervisors and middle managers are critical to successful adoption. They translate governance and change messages into daily priorities. If they are unconvinced or underprepared, AI governance will falter regardless of top-level support.

Support them by:

- Providing clear, manager-focused summaries of AI systems: purpose, risk tier, oversight mode, and key dos/don'ts.

- Equipping them with talking points for team discussions about AI and governance.

- Involving them in pilot phases so their feedback informs adjustments.

- Giving them metrics that include both performance and responsible use (e.g., override rates, incident reports), not just productivity.

When supervisors are confident and informed, they can coach teams through the inevitable ambiguities of early AI adoption.

14.9   Embedding AI governance into daily routines

To make governance and adoption durable, integrate AI-related practices into existing routines rather than creating entirely new ones:

- Team stand-ups
  Include short check-ins on AI behavior: any odd outputs, pain points, or positive impacts.

- 1:1s and performance discussions
  Ask how individuals are using AI, where they feel confident or uncertain, and whether they understand their oversight role.

- Operational reviews
  Discuss AI metrics alongside traditional KPIs: not just volume and speed, but incidents, overrides, and user feedback.

- Retrospectives and lessons-learned sessions
  For projects involving AI, explicitly ask what was

learned about governance, adoption, and user experience.

These practices normalize governance conversations and prevent AI from being treated as a mysterious side topic.

Diagram 14.4: Governance in Daily Workflows

14.10 Handling resistance: listening and adjusting without abandoning governance

Resistance to AI and governance can signal deeper issues, such as unrealistic expectations, flawed workflow design, or insufficient communication. Instead of dismissing resistance, managers should:

- Listen for patterns
  Are multiple people raising the same concern (e.g.,

AI suggestions are frequently off in a particular scenario)?

- Separate resistance to AI from resistance to poor design
Some "AI resistance" is actually resistance to added clicks, confusing interfaces, or unclear rules.

- Adjust what can be improved
Simplify workflows, tweak configurations, refine training, or adjust thresholds based on real-world experience.

- Hold firm where governance is non-negotiable
For example, fairness checks, incident reporting, or restrictions on unapproved uses should not be relaxed to reduce friction.

The message to teams should be: governance is here to stay, but **how** we implement it and **how** AI fits into your work is open to improvement.

14.11 The manager's change-and-adoption checklist for AI-governed systems

Use this checklist before and after deploying a governed AI system:

Before deployment:

- Have we identified all key stakeholder groups and their concerns?

- Do we have clear messages about what is changing, what is not, and why?

- Does training cover governance, oversight, and escalation—not just "how to use the tool"?

- Do supervisors understand their role in reinforcing responsible use?

After deployment:

- Are we monitoring both system metrics and human adoption indicators (usage rates, override patterns, user feedback)?

- Do people know how to report issues or concerns, and do they see responses?

- Are we using feedback to adjust workflows, training, and configuration while maintaining core governance requirements?

If you cannot answer "yes" to most of these questions, your change and adoption plan is likely underpowered relative to the governance expectations you have set.

14.12 AI Reality Check: adoption takes time and iteration

Even with strong change management, adoption is rarely instantaneous. People need time to understand new tools, see them perform, and trust that governance protects them as well as the organization. Expect:

- A learning curve with initial dips in speed or comfort.

- Mixed reactions across teams and individuals.

- The need for multiple rounds of communication and training.

The goal is not to force immediate, unquestioning adoption, but to build **sustained, responsible use**. That requires patience, reinforcement, and a willingness to evolve both AI systems and governance practices in response to experience.

14.13 How this chapter advances your governance journey

In this chapter, you connected AI governance to the human realities of change and adoption. You learned to map stakeholders, design change with governance in mind, train for oversight and accountability, shape trust, support supervisors, and embed governance into daily routines. You also gained practical checklists to ensure that when AI systems go live, people are ready to use them responsibly— not just technically able to access them.

Next, we turn outward to your broader ecosystem. In Chapter 15, we will explore **Vendor and Ecosystem Governance for AI**—how to extend your governance expectations beyond your own walls to the partners, platforms, and providers that power much of modern AI.

15 Vendor and Ecosystem Governance for AI

15.1 Opening vignette: "We outsourced the capability, not the accountability."

A public sector IT manager proudly reported that their new AI-powered eligibility assistant was "fully managed" by a major vendor. The system classified applications and suggested eligibility decisions, integrated into the agency's portal via an API. When questions arose about fairness and transparency, the manager reassured leadership: "The vendor is certified; they handle all that." Months later, advocacy groups challenged specific denials and demanded explanations. The vendor provided a generic white paper on their platform, but could not explain how the model used this agency's data in context—or how decisions aligned with local policy and law. Regulators made it clear: "You remain accountable for how AI is used in your services." The lesson was painful but clear: you can outsource AI capabilities, but not **AI accountability**. This chapter explains how to govern AI across vendors and your wider ecosystem.

15.2 Why ecosystem governance matters in AI

Modern AI rarely lives entirely inside your data center. It emerges from an ecosystem:

- Cloud providers host your data and models.

- SaaS vendors embed AI into their products.

- Niche AI firms provide specialized models and tools.

- Integrators and consultants stitch it all together.

Each partner influences the behavior, risk, and compliance posture of your AI systems. If you treat each relationship purely as a procurement transaction, you risk importing:

- Opaque models you cannot explain.

- Data flows you cannot fully trace.

- Dependencies you cannot easily manage or replace.

Ecosystem governance is about extending your AI governance expectations—on risk, data, ethics, and accountability—to the vendors and partners you rely on. It ensures that your external relationships support, rather than undermine, your internal governance.

15.3 Defining vendor and ecosystem governance for AI

Vendor and ecosystem governance for AI consists of:

- Criteria for selecting AI-capable providers.

- Contractual and policy requirements related to AI usage.

- Onboarding processes that configure AI services within your governance bounds.

- Ongoing monitoring of vendor performance, risk, and incidents.

- Clear exit and transition strategies for AI-dependent services.

It is not a separate governance system. It is your existing AI and vendor governance, extended to cover AI-specific risks and responsibilities in the ecosystem.

You, Your AI Governance, and the Ecosystem

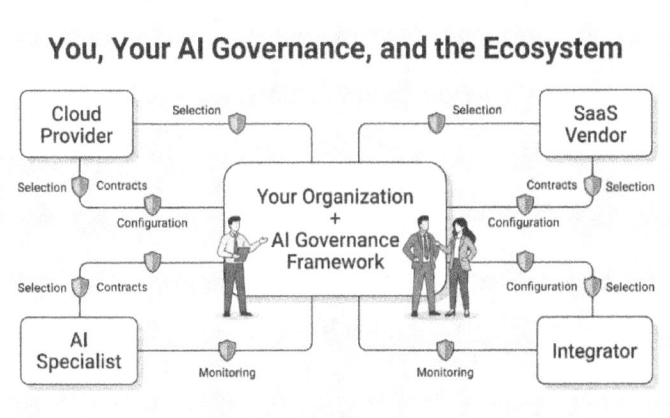

15.3 AI-aware vendor selection: beyond features and price

Traditional vendor selection tends to focus on functionality, cost, and basic security. For AI, selection must also take governance and responsibility into account. Practical AI-aware criteria include:

- Transparency and documentation
  Can the vendor describe, in understandable terms, what their AI does, how it is trained, and its known limitations?

- Data handling and privacy practices
  How do they process, store, and protect your data?
  Do they reuse it to train other models? Can they
  meet your residency and retention requirements?

- Risk and ethics posture
  Do they have their own AI governance principles,
  risk assessments, and fairness practices? Can they
  share evidence (summaries, certifications) of these?

- Control and configuration options
  Do they provide controls you can use to enforce
  your policies (e.g., content filters, logging, opt-out
  of data retention)?

- Support for monitoring and audit
  Do they provide logs, metrics, and interfaces to
  monitor behavior and support audits?

These criteria should be built into your vendor evaluation
templates, not added ad hoc.

15.4 Contracting as a governance tool for AI

Contracts are one of your strongest governance levers in the
ecosystem. For AI-related services, contracts should
address:

- Data use and ownership
  Who owns the data and derived models? Can the
  vendor reuse your data, and under what conditions?
  How is data deleted at the end of the contract?

- Transparency obligations
  What documentation must the vendor provide about models, updates, and known limitations? How quickly must they inform you of material changes?

- Security and compliance requirements
  Which standards and certifications must they maintain? How will they demonstrate compliance?

- Incident notification and cooperation
  How quickly must they notify you of AI-related incidents (e.g., data leakage via AI tools, misbehavior)? What cooperation will they provide in investigations?

- Performance and fairness commitments (where feasible)
  For high-risk use cases, can they commit to certain testing practices or assist in fairness and robustness assessments?

- Audit and assessment rights
  Can you, or a trusted third party, periodically review aspects of their AI practices relevant to your risk exposure?

Well-crafted terms align external behavior with your internal governance framework and clarify expectations before something goes wrong.

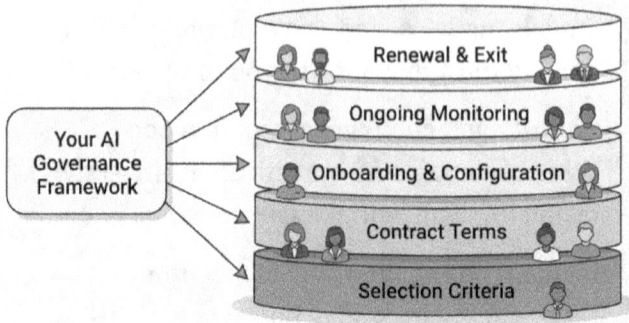

Vendor Governance Stack

15.5   Onboarding vendors into your AI governance framework

Once selected and contracted, AI vendors must be onboarded into your governance environment:

- Register vendor AI services in your AI portfolio and system inventory.

- Classify each vendor-supported use case by risk tier and autonomy level.

- Define internal owners for vendor services (business and technical).

- Configure the service according to your policies (e.g., logging enabled, data minimization, restricted features).

- Document integration points: what data flows where, what actions the AI can trigger.

Onboarding is your chance to ensure the vendor's AI does not bypass your controls. It is where you connect the vendor's capabilities to your **AI operating model**, risk framework, and data governance.

15.6   Managing multi-vendor and multi-cloud complexity

Most organizations will not have a single AI vendor or cloud provider. Over time, you may accumulate:

- Multiple clouds hosting different data and models.

- SaaS platforms with embedded AI for CRM, HR, finance, and more.

- Specialized vendors for domain-specific predictions or generative tasks.

This complexity increases the risk of:

- Inconsistent controls and configurations.

- Overlapping or conflicting AI behaviors.

- Difficulty tracing responsibility when incidents involve multiple providers.

To manage this:

- Define baseline AI governance requirements that apply to all vendors (e.g., data handling minimums, logging, documentation).

- Maintain a central registry of AI-relevant vendors and services.

- Align configuration guidelines across platforms (e.g., common standards for what data can enter generative services).

- Use architecture and security review boards to ensure integrations respect cross-vendor patterns and constraints.

The aim is not uniformity for its own sake, but sufficient consistency to make oversight manageable and responsibility traceable.

15.7   Ecosystem risk: concentration, dependency, and cascading failures

AI ecosystem risk is not only about individual vendors; it is also about how they interact:

- Concentration risk
  Relying heavily on a small number of providers for critical AI capabilities (e.g., a single cloud model for all high-stakes decisions).

- Dependency risk
  Integrations are so tightly coupled that vendor changes or outages cascade through your workflows.

- Model monoculture
  Many systems relying on the same underlying model, making systemic biases or failures more impactful.

Governance should:

- Identify critical dependencies and concentrations in your AI portfolio.

- Plan for contingencies (e.g., fallbacks, alternative providers, partial manual operation) if key services become unavailable or unsuitable.

- Periodically question whether your ecosystem structure still aligns with your risk appetite and regulatory expectations.

This is particularly important for public services, safety-critical functions, and high-impact financial or healthcare decisions.

Ecosystem Risk View

15.8   Monitoring vendor AI performance and behavior

Ongoing governance requires you to monitor vendor-provided AI services, not just your in-house systems. Practical steps:

- Incorporate vendor services into your monitoring dashboards where possible.

- Track service availability, latency, and error rates from your perspective.

- Monitor AI outputs for unexpected changes in patterns, even if vendor assures stability.

- Track incidents and near-misses involving vendor AI, including user complaints and escalation.

- Schedule periodic review meetings with key vendors to discuss performance, incidents, and roadmap changes.

Where vendors provide their own dashboards or reports, integrate those insights into your governance forums rather than letting them sit in separate silos.

15.9   Aligning incident response with vendors

AI incidents often involve vendor components. You need a coordinated incident response approach:

- Clear contact paths
  Know exactly how to reach vendor incident teams
  and what response times to expect.

- Shared definitions
  Agree on what qualifies as an AI-related incident
  and how severity is classified.

- Joint investigation
  Ensure you can access the logs, configurations, and
  model details needed to understand what happened.

- Coordinated communication
  Align internal and external messaging, recognizing
  that both your organization and the vendor have
  reputational stakes.

- Post-incident improvement
  Update your configurations, policies, and
  sometimes contracts based on what you learn
  together.

This alignment turns vendors into partners in resilience
rather than mere vendors in a crisis.

15.10 Integrating ecosystem governance into your internal structures

Vendor and ecosystem governance must be anchored in your
internal processes:

- Procurement and legal
  Use standardized AI governance criteria and contract clauses as starting points, with variation only when justified.

- Architecture and security review boards
  Evaluate vendor AI integrations against your technical and risk standards.

- AI governance council
  Review high-risk vendor proposals and monitor the AI ecosystem as part of portfolio and risk oversight.

- Risk and compliance functions
  Include vendor AI in risk assessments, audits, and regulatory reporting.

This integration ensures that external capabilities remain within your governance perimeter instead of forming islands of unmanaged risk.

15.11 The manager's checklist: ecosystem and vendor governance readiness

Use this checklist to gauge your current state:

- Do we know which vendors and platforms provide AI capabilities in our environment?

- Do our vendor selection and contract processes include AI-specific governance and risk requirements?

- Are vendor AI services registered in our AI portfolio and risk frameworks with clear owners?

- Do we have visibility into vendor AI behavior through logs, metrics, or agreed reports?

- Are AI-related incidents involving vendors covered by joint response plans and clear escalation paths?

If the answer to several of these is "not yet," strengthening vendor and ecosystem governance is a high-priority step in maturing your overall AI governance posture.

15.12 AI Reality Check: You can share responsibility, but not accountability

It is tempting to assume that large vendors and platforms "have governance covered." They often do have robust internal practices—but those practices are designed for their obligations and risk appetite, not yours. Your context, stakeholders, and regulations may differ significantly.

In practice, you can **share responsibility** with vendors for specific tasks—security controls, certain aspects of testing, or tooling—but you cannot outsource **accountability** to your leadership, regulators, or the public. When AI goes wrong in your services, you will still be called to explain, defend, and remediate. Effective vendor and ecosystem governance is how you ensure that when you stand up in those moments, you have a clear story, evidence of due diligence, and partners ready to support—not hinder—your response.

15.13 How this chapter advances your governance journey

In this chapter, you extended your AI governance outward into the ecosystem of vendors and partners that power modern AI. You learned how to incorporate AI-specific criteria into vendor selection and contracting, onboard services into your governance framework, manage multi-vendor complexity and ecosystem risk, monitor vendor AI behavior, and coordinate incident response. You also reinforced the central truth that accountability for AI remains with you, even when capabilities come from outside.

In the final additional chapter, **Chapter 16 – Crisis Management and Incident Response for AI Systems**, we will address what happens when governance is tested under pressure. You will learn how to define AI incidents, integrate AI into your incident response playbooks, and turn crises into catalysts for better governance across your portfolio.

16 Crisis Management and Incident Response for AI Systems

16.1 Opening vignette: "The alert no one owned."

An AI-powered risk scoring system had been running quietly in production for nearly a year. It flagged cases for additional review in a high-stakes program. One morning, analysts noticed something strange: the system's recommendations had shifted overnight, suddenly downgrading many cases from "high" to "low" risk. A junior analyst raised a concern to her supervisor, who suggested it might just be "a fluctuation" and told her to keep working. Days later, a journalist published an investigation suggesting that high-risk cases were being missed. Leadership scrambled: Was the AI at fault? Was it a data issue? Had something changed in the model? No one knew exactly who owned incident response for AI. Logs were fragmented, responsibilities unclear, and communication reactive and defensive. This chapter is about preventing that situation by building a **crisis management and incident response** capability tailored to AI systems.

16.2 Why AI incidents need special attention

Your organization likely already has incident response processes in place for cybersecurity incidents, outages, or data breaches. AI introduces additional incident types and patterns that those processes do not fully cover:

- Behavioral incidents
  The AI behaves unexpectedly: sudden drift, biased outputs, hallucinations, or unsafe actions.

- Content incidents
  Generative systems produce harmful, inappropriate, or misleading content that reaches users.

- Decision incidents
  AI-supported decisions systematically disadvantage groups, violate policy, or create safety risks.

- Data-related AI incidents
  Sensitive data appears in AI outputs or is used in ways stakeholders did not consent to, especially through generative tools.

These incidents may not involve a system "down" or a classic data breach. The system might be up and secure, but it is making **bad or misaligned decisions**. Existing incident processes must be expanded to handle this new class of risk.

16.3 Defining what counts as an AI incident

To respond effectively, you must clearly define what constitutes an **AI incident** in your organization. A practical definition:

An AI incident is any event where an AI system's behavior, outputs, or use deviates from intended, governed operation in a way that could cause or has caused harm, unfairness, non-compliance, or significant loss of trust.

You can categorize AI incidents by type and severity:

- Types
    - Performance/behavioral (e.g., unexplained drift, wrong outputs at scale)
    - Fairness/ethics (e.g., systematic disadvantage to a group)
    - Safety/critical decision failures
    - Security/misuse (e.g., prompt injection, model abuse)
    - Privacy/data leakage via AI interactions
    - Reputational (e.g., harmful generated content reaching the public)
- Severity
    - Low: limited scope, easy to contain, minimal impact.

- Medium: broader scope, noticeable impact, manageable internally.

- High: significant harm or exposure, likely external attention.

- Critical: safety, legal, or systemic implications; requires executive and possibly regulatory involvement.

These definitions should be documented and socialized so people know when to escalate from "issue" to "incident."

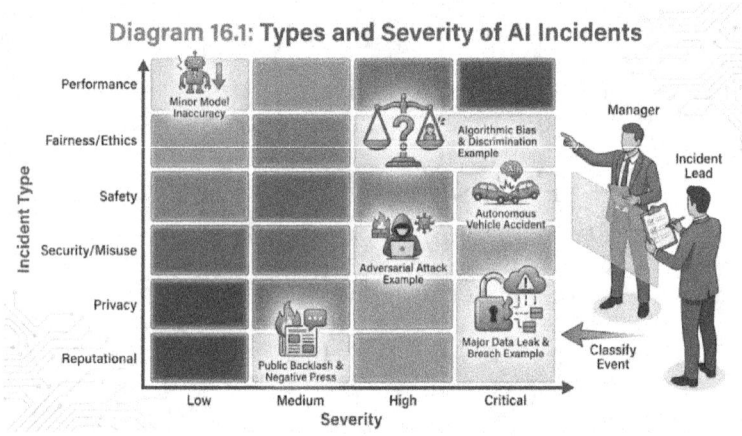

Diagram 16.1: Types and Severity of AI Incidents

16.4   Integrating AI into your existing incident response structure

Rather than building a separate AI-only incident process, integrate AI into your existing incident response framework, adding AI-specific steps and roles:

- Detection
  Issues can be detected via monitoring dashboards, user reports, audits, or external complaints. For AI, detection must include behavioral and fairness signals, not just uptime and errors.

- Triage
  First-line responders assess whether the issue meets the AI incident criteria and assign a preliminary severity rating.

- Response team activation
  For medium and above, a defined AI incident response team is activated, including technical, data, risk, and business roles.

- Investigation and containment
  The team works to understand root causes and contain further harm (e.g., rolling back a model version, restricting access, or switching to manual review).

- Communication
  Internal stakeholders are informed promptly; external communication is planned based on severity and obligations.

- Recovery and improvement
  Systems are restored to safe operation, and governance improvements are identified and implemented.

The key is to ensure your existing structure recognizes AI as a **special class of incident** with specific expertise and steps, not an undefined edge case.

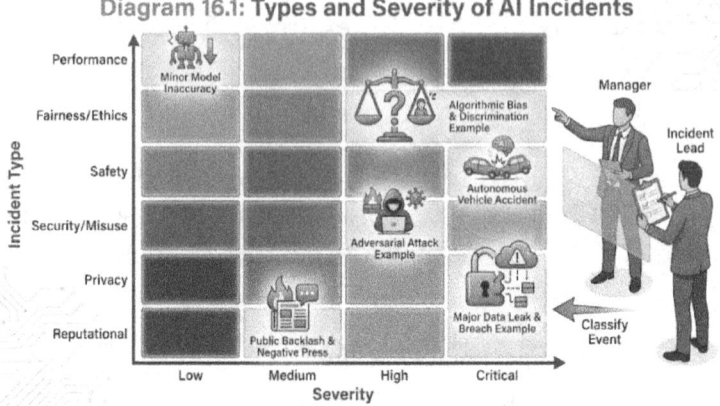

Diagram 16.1: **Types and Severity of AI Incidents**

16.5 Roles in AI incident response

AI incidents require a multi-disciplinary response team. Typical roles include:

- AI incident lead
  Coordinates the response, ensures steps are followed, and acts as the primary point of contact. Often, a senior manager has both technical and governance understanding.

- Technical lead (AI/IT)
  Diagnoses system behavior, inspects logs, and executes technical containment (e.g., rolling back models, changing configurations).

- Data and model expert
  Understands the data pipelines and models involved; investigates potential drift, data issues, or unexpected inputs.

- Business/mission owner
  Evaluates operational and stakeholder impact;
  decides on temporary process changes (e.g., manual
  overrides, priority shifts).

- Risk/compliance/legal
  Assesses regulatory, contractual, and policy
  implications; advises on notification and
  remediation obligations.

- Communications
  Manages internal and external messaging to ensure
  accuracy, consistency, and alignment with
  governance principles.

These roles can overlap in smaller organizations, but each
responsibility must be clearly assigned in incident
playbooks.

16.6   Defining AI incident detection channels

Effective incident response depends on **how issues surface**.
For AI, detection channels include:

- Automated monitoring
  Dashboards flag anomalies such as sudden changes
  in output distributions, error rates, drift metrics, or
  unusual usage patterns.

- Frontline user reports
  Staff interacting with AI (e.g., agents, analysts,

clinicians) notice odd outputs or patterns and report them via defined channels.

- Stakeholder complaints
Citizens, customers, or employees raise concerns about AI-related decisions or content.

- Internal audits and reviews
Periodic checks identify compliance or fairness issues.

- External signals
Media coverage, regulator questions, or vendor notifications indicate potential problems.

You should:

- Define specific channels for AI-related concerns (e.g., tagged incident categories in ticketing systems).

- Make it clear that raising AI concerns is expected and valued.

- Ensure reports are triaged promptly, not lost among generic system tickets.

16.7 Temporary safeguards: stabilizing in a crisis

When an AI incident is confirmed, you need **temporary safeguards** to reduce harm while investigating. Options include:

- Model rollback
  Revert to a previous model version known to behave acceptably.

- Scope reduction
  Limit the AI system's use to lower-risk contexts or smaller user groups.

- Human override elevation
  Require additional human review or approvals for AI-influenced decisions.

- Feature disablement
  Temporarily turn off specific AI features (e.g., generative responses to external users).

- Manual fallback
  Revert to manual processes for critical decisions until the issue is understood.

Your incident playbooks should predefine which safeguards are appropriate at different severity levels, enabling quick, consistent decisions.

16.8   Communication during AI incidents: internal and external

Communication can either calm or inflame a crisis. For AI incidents, communication plans should address:

- Internal communication

- Inform relevant teams quickly: what happened, what is known, what has been done so far.

- Guide interim operational changes and talking points for frontline staff.

- Avoid blame; focus on facts, safety, and next steps.

- External communication (when needed)

  - Be transparent about the issue in terms appropriate to the audience (avoid jargon, be honest about uncertainty).

  - Explain what you are doing to mitigate harm and prevent recurrence.

  - Align messages with your Responsible AI commitments and governance principles.

Risk, legal, and communications roles must collaborate closely so statements are accurate, appropriately scoped, and consistent with your obligations.

16.9   Post-incident review: turning crises into governance improvements

Once an AI incident is contained, a structured post-incident review (sometimes called a "post-mortem" or "lessons learned" session) should address:

- What happened?
  Timeline of events, system behavior, detection method, and stakeholder impact.

- Why did it happen?
  Root causes: data issues, model design, integration, governance gaps, cultural factors (e.g., unreported concerns).

- What worked and what did not in the response?
  Speed of detection, clarity of roles, effectiveness of safeguards, and quality of communication.

- What will we change?

  - Technical changes: data, models, thresholds, configurations.

  - Governance changes: risk tiers, policies, review processes, and monitoring standards.

  - Cultural and training changes: reinforcing escalation norms or oversight expectations.

The review should be documented and fed back into your governance framework—updating templates, standards, and training so the same issue is less likely to recur.

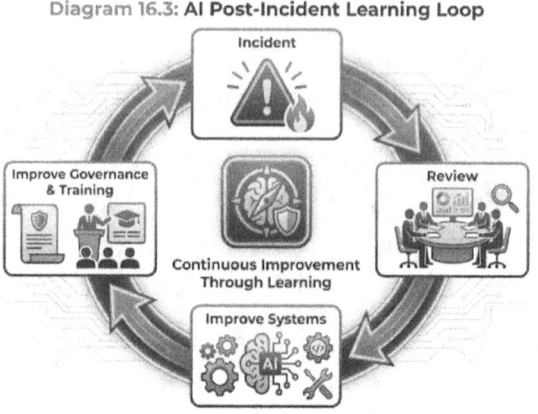

Diagram 16.3: AI Post-Incident Learning Loop

16.10 Aligning AI incident response with your risk and governance framework

AI incidents should not sit outside your broader risk and governance structures. To align them:

- Include AI incidents in your enterprise risk registers and reports.

- Ensure severe AI incidents are reviewed by your AI governance council and, where appropriate, risk committees and senior leadership.

- Treat repeated incidents in a particular domain as signals that governance, data, or culture requires deeper intervention.

- Use incident data to refine your risk tiering, testing standards, and monitoring baselines.

This alignment ensures that crises are not just "put out" but are recognized as **feedback** on the adequacy of governance and operating model design.

16.11 The manager's checklist: AI incident readiness

Use this checklist to gauge how prepared you are:

- Do we have a clear, documented definition of AI incidents (types and severities)?

- Are AI-specific steps integrated into our existing incident response processes and runbooks?

- Do we have named roles (even if part-time) for AI incident lead, data/model expert, and business owner?

- Are detection channels for AI issues defined and known to frontline staff?

- Do we have predefined temporary safeguards and escalation paths for high-severity AI incidents?

- Do we conduct post-incident reviews that feed into governance and training improvements?

If the answer to several is "no" or "not sure," AI incident readiness should be a priority in your governance roadmap.

16.12 AI Reality Check: the goal is resilience, not zero incidents

No matter how strong your governance, incidents will occur. Data will change, models will drift, unexpected inputs will surface, and edge cases will slip through. The goal of crisis management and incident response is not to guarantee that nothing ever goes wrong—that is impossible. The goal is **resilience**:

- Detecting issues early.

- Containing harm quickly.

- Responding transparently and professionally.

- Learning systematically and improving over time.

Organizations that handle AI incidents well often strengthen trust rather than lose it. Stakeholders see that you take responsibility seriously and that your governance is more than words on paper.

16.16 How this chapter completes your governance journey

In this chapter, you prepared for the moments when AI governance is tested under pressure. You learned how to define AI incidents, integrate them into incident response, assign roles, deploy temporary safeguards, communicate effectively, and use crises as catalysts for improvement.

Together with the earlier chapters—defining AI governance, building an operating model, managing data integrity, addressing risk and ethics, governing cloud and generative AI, shaping culture, managing portfolios, working with vendors, and measuring maturity—you now have a comprehensive, manager-ready guide to AI and IT governance in a modern enterprise.

17 Conclusion

This book began with a simple observation: AI is no longer an experiment at the edges of IT. It is embedded in systems that decide, workflows that shape lives, and platforms that operate across clouds, vendors, and jurisdictions. For a serious manager, that reality creates both urgency and opportunity. Urgency, because unmanaged AI can amplify risk silently and at scale. Opportunity, because managers who build strong AI governance now will define how their organizations—and in many cases their communities— experience AI for years to come.

Across these chapters, you have walked a deliberate path from concepts to practice. You started by reframing AI governance not as a compliance add-on, but as the core discipline for managing systems that decide. You saw how traditional IT governance remains necessary but incomplete when models learn, drift, and act in complex environments. You explored the four recurring dimensions—value, risk, control, and trust—and learned to apply them as a simple, repeatable lens to every AI conversation you lead.

From there, you moved into the realities of operating. You designed an AI operating model that makes "how we do AI here" explicit: who identifies opportunities, how they are evaluated, who builds and validates models, who approves deployment, and who monitors behavior over time. You saw that the operating model is not a diagram; it is the lived pattern of roles, processes, and decisions that either supports or undermines your governance ambitions. By defining roles, clarifying decision rights, and embedding "responsible by design" checkpoints into the lifecycle, you created a structure that can scale beyond individual champions or one-off projects.

You then built out the governance framework itself: principles, policies, roles, risk tiers, and lifecycle processes that sit atop the operating model and give it authority. You learned to right-size governance using risk tiers and autonomy levels, applying heavier oversight where the stakes are highest and lighter processes where risk is low but learning is valuable. You saw that algorithmic bias, model drift, and opacity are not niche technical issues but central governance drivers that must be reflected in risk assessments, approvals, and monitoring.

The middle of the book went deep into the elements that make AI governance real. You anchored data at the heart of the story, recognizing that data integrity and accountability are non-negotiable foundations for any responsible AI program. You linked AI governance to data governance, defined AI-specific data quality and lineage needs, and built a lifecycle perspective that covers data from collection to archival. You integrated AI risk into enterprise risk management, treating it as a first-class citizen alongside financial, operational, and security risks. You translated Responsible AI principles into concrete practices: fairness tests, oversight patterns, transparency for stakeholders, and clear documentation.

You also moved beyond technical systems into human systems. You operationalized ethics and transparency by connecting principles to standards, workflows, and artifacts. You designed governance for cloud and hybrid environments, recognizing that while providers share responsibility, accountability remains with you. You tackled generative and autonomous systems by defining levels of autonomy, guardrails, and oversight patterns to keep powerful tools within acceptable boundaries. You invested in people, skills, and culture, accepting that governance lives or dies in the daily decisions of managers, teams, and frontline staff.

In the final chapters, you zoomed out to the portfolio and ecosystem. You learned to manage AI as a portfolio— prioritizing, sequencing, and sometimes stopping initiatives based on value, risk, and readiness. You extended governance into your vendor landscape, treating contracts, selection, configuration, and ongoing review as governance levers rather than just procurement steps. You built measurement and maturity into your approach, turning governance from a static design into a learning system that can adjust as your context changes. And you prepared for the inevitable: AI incidents and crises. By defining AI incidents, integrating them into response playbooks, and treating them as sources of improvement rather than as sources of embarrassment, you created resilience rather than chasing an impossible zero-incident ideal.

Taken together, these elements form a coherent playbook for AI-modern IT. They equip you to move beyond slogans and point solutions. Instead of "doing AI" in a scattershot way, you can lead AI as part of a governed, mission-aligned transformation. Instead of being surprised by AI risks, you can anticipate and channel them. Instead of relying on individual heroics, you can build an operating model and culture where responsible AI is how work gets done.

Throughout this book, it has taken a manager's-first perspective. You are not expected to tune models or design deep learning architectures. You are expected to ask the right questions, insist on the right evidence, and make the right trade-offs. You are accountable for connecting AI to mission, ensuring that governance is proportional and real, and shaping a culture where people feel both empowered to use AI and obligated to use it responsibly. That is not about saying "no" to AI. It is about saying "yes, under these conditions" and being able to explain those conditions to anyone who asks.

As you close this book, several practical commitments can help you turn its ideas into durable practice:

- Treat every new AI proposal as an opportunity to apply the full governance lens: value, risk, control, and trust, not just technical feasibility.

- Make your AI operating model explicit and visible, so teams know where AI work belongs, how it moves through the lifecycle, and who owns it at each stage.

- Keep your AI portfolio current and reviewed, so leadership sees where AI is delivering, where it is stuck, and where governance or data investments are needed.

- Invest in people: equip managers and staff with governance literacy, make champions visible, and align incentives with responsible behavior.

- Accept that your framework is a starting point, not an endpoint. Use incidents, audits, and feedback to regularly update policies, templates, and practices.

Most importantly, remember that AI governance is not a separate agenda competing with modernization. It is the way you do modernization. It is how you scale AI without losing control, how you build trust without halting innovation, and how you honor the responsibilities that come with deploying systems that decide in environments that matter.

You will face pressure—from vendors, from internal champions, from headlines—to move faster. Sometimes that pressure will be justified; sometimes it will tempt you to cut corners. In those moments, your role as a serious AI manager comes into sharp focus. You are the one who can say: "We will move quickly, but we will move with governance. We will use AI, but we will use it in ways we can explain, defend, and improve." That stance will not always be easy. But it is how you protect your mission, your stakeholders, and your teams in an AI-driven world.

You now have the frameworks, language, and practical tools to lead that work. The rest is leadership: the daily choices to apply what you know, to bring others along, and to keep governance and responsibility at the center of AI-modern IT.

Works Referenced

Alberts, Ron, et al. "Strengthening AI Governance: International Policy Frameworks and the GIRAI Model." *ITU Journal on Future and Evolving Technologies*, vol. 6, no. 3, Sept. 2025, pp. 271–289.[itu]

All Tech Is Human. *Responsible AI Impact Report: Urgent Risks, Emerging Safeguards, and Public Interests.* All Tech Is Human, 1 Dec. 2025.[alltechishuman]

Angle, Kevin, and Rachel Marmor. "USA: AI Governance and the Law in 2025." *OneTrust DataGuidance*, 4 May 2025.[hklaw]

Cimplifi. *The Updated State of AI Regulations for 2025.* Cimplifi, 29 Apr. 2025.[cimplifi]

European Commission. *Regulation (EU) 2024/1689 Laying Down Harmonised Rules on Artificial Intelligence (Artificial Intelligence Act). Official Journal of the European Union*, 2024.[cimplifi]

Future of Life Institute. *2025 AI Safety Index: Summer 2025 Edition.* Future of Life Institute, 13 Jan. 2026.[futureoflife]

Google. *Responsible AI Progress Report.* Google, Feb. 2025.[ai]

Holland & Knight. "Global AI Governance: Five Key Frameworks Explained." *The AI Journal*, 13 Aug. 2025.[bradley]

Hovy, Dirk, and Isabelle Augenstein. "AI Risk Management in Practice: Operationalizing the NIST AI RMF in Large Organizations." *AI and Ethics*, vol. 5, no. 2, 2025, pp. 145–162.[sciencedirect]

International Telecommunication Union. *Strengthening AI Governance: International Policy Frameworks and National Readiness*. ITU, 2025.[itu]

Jaume-Palasí, Lorena, et al. *Implementing the EU AI Act: A Practical Guide for Public Sector Organizations*. World Economic Forum, 2024.[winmarkglobal]

Krafft, P. M., et al. "Algorithmic Governance and the Limits of Formalization." *Annual Review of Law and Social Science*, vol. 20, 2024, pp. 311–332.[sciencedirect]

Liu, Han, and Maria Elvira. "Responsible Artificial Intelligence Governance: A Review and Research Agenda." *Information & Management*, vol. 61, no. 3, 2024, article 103012.[sciencedirect]

McDermott, Orla, et al. "AI Governance and the Medical Device Lifecycle: Lessons from Industry 4.0." *Sustainability*, vol. 16, no. 2, 2024, article 14650.[ppl-ai-file-upload.s3.amazonaws]

National Institute of Standards and Technology. *Artificial Intelligence Risk Management Framework (AI RMF 1.0)*. NIST, Jan. 2023.[ppl-ai-file-upload.s3.amazonaws]

National Law Review. "Artificial Intelligence Legislative Update: Emerging US State and Federal AI

Laws." *The National Law Review*, 13 Jan. 2026.[natlawreview]

Organisation for Economic Co-operation and Development. *OECD Framework for the Classification of AI Systems*. OECD, 2024.[sciencedirect]

Organisation for Economic Co-operation and Development. *OECD Principles on Artificial Intelligence*. OECD, 2019.[ppl-ai-file-upload.s3.amazonaws]

Pauwels, Eleonore, and David Leslie. *Operationalising Responsible AI: From Principles to Practice in Public Institutions*. World Economic Forum, 2023.[winmarkglobal]

Pieters, D. P. "A Decision Support System for Selecting IT Audit Areas Using a Capital Budgeting Approach." 2015.[ppl-ai-file-upload.s3.amazonaws]

Sato, Keiko, et al. "Developing an AI Governance Framework for Safe and Responsible Use of AI in Health Care: Protocol for a Multimethod Study." *JMIR Research Protocols*, vol. 14, no. 1, 2025, article e75702.[researchprotocols]

Sheikh, Nida. *AI Governance and Frameworks: How to Manage AI Risks and Compliance*. PM World Journal, vol. 14, no. 7, July 2025.[pmworldlibrary]

Smith, John, et al. "Enterprise AI Governance: Structures, Roles, and Processes in Global Firms." *MIS Quarterly Executive*, vol. 23, no. 4, 2024, pp. 211–228.[sciencedirect]

United Nations Educational, Scientific and Cultural Organization. *Recommendation on the Ethics of Artificial Intelligence: Implementation Guidelines 2023–2025.* UNESCO, 2023.[ai21]

U.S. Executive Office of the President. *Executive Order on Safe, Secure, and Trustworthy Development and Use of Artificial Intelligence.* The White House, 30 Oct. 2023.[cimplifi]

U.S. Federal Trade Commission. *Aiming for Truth, Fairness, and Equity in Your Company's Use of AI.* FTC, 19 Apr. 2021.[ppl-ai-file-upload.s3.amazonaws]

Winmark. *AI Governance: Frameworks, Best Practices and Policies in 2024.* Winmark Global, 2024.[winmarkglobal]

World Economic Forum. *Global Artificial Intelligence Governance: Balancing Innovation and Responsibility.* World Economic Forum, 2023.[ppl-ai-file-upload.s3.amazonaws]

World Economic Forum. *AI Governance Alliance: Global Guidelines for Responsible AI Systems.* World Economic Forum, 2024.[winmarkglobal]

World Health Organization. *Ethics and Governance of Artificial Intelligence for Health: Policy and Governance Considerations 2023–2025.* WHO, 2023.[researchprotocols]

www.ingramcontent.com/pod-product-compliance
Lightning Source LLC
Chambersburg PA
CBHW051306220526
45468CB00004B/1230